POW!
The Planning of Wetlands
An Educator's Guide

Karen L. Ripple and Edgar W. Garbisch

Environmental Concern Inc.
St. Michaels, Maryland 21663

Published by Environmental Concern Inc.
P. O. Box P, 201 Boundary Lane
St. Michaels, Maryland 21663
410-745-9620, www.wetland.org

Environmental Concern Inc. is a nonprofit corporation devoted to wetland education, wetland research, and development and application of technology in the restoration and construction of wetlands.

Printed in the United States of America
on recycled paper.

Library of Congress Cataloging-in-Publication Data

Ripple, Karen L., 1946-
POW! : the planning of wetlands : an educator's guide / Karen L. Ripple and Edgar W. Garbisch
p. cm.
Includes bibliographical references (p.).
ISBN 1-883226-05-8 (pbk.)
1. Wetlands. 2. Wetlands--Study and teaching (Secondary)--Activity programs. 3. Constructed wetlands--Design and construction. I. Garbisch, Edgar William. II. Title.

QH87.3 .R56 2001
578.768'071'2--dc21

2001022927

ACKNOWLEDGMENTS

I am indebted to Dr. Edgar Garbisch, "the marsh builder," not only for writing *Part I: Background Information*, but also for allowing me the freedom to invest so much time and effort in a project with very little outside funding. The process of writing and testing the activities, then of bringing everything together as a publication has been immensely educational and challenging. Throughout this project Dr. Garbisch has patiently answered technical questions, explained wetland processes, and never wavered in his support and belief that this project would have a successful conclusion. Thank you for letting me try my wings.

Special thanks go to Dr. Mark Kraus for writing the manuscript that evolved into *Part I*. Although the original manuscript was never fully developed or printed, it did serve as a guide for background portions of this new publication. Thank you, Meredith E. Lathbury, Esq., for writing the heart of the chapter *Permits* and helping to start this entire project. Thanks also to two extraordinary young women, Renee Blacken and Lillian Harris, for their assistance in creating the *Glossary*.

Special thanks to Joanna Garbisch and Suzanne Pittenger-Slear for the creative but often tedious work of formatting and layout of the book and cover, and Barbara Jarjoura for helping with that process. For the beautiful artwork gracing the pages of this book, thank you, Carrie Barcomb, Dawn Biggs, Jennifer Bittorf, J. Robert Gorski, William C. Gorski, Fred Hartman, Deborah Johnson, Elsa Naser, Britt Eckhardt Slattery, Renée F. Wilson, and the University of Florida Center for Aquatic Plants (Gainesville). A very special thank you goes to Laura Malick for her invaluable assistance in proofreading and editing.

For assistance with the first two pilot workshops and for helping to create some of the initial activities, thank you, Deborah Herr-Cornwell. I also wish to thank Albert McCullough III, Doreen Dudek, Sally Naser, and Leslie Hunter for technical assistance and support. David Christopher, thank you for being my educational sounding board and providing the computer expertise that eased the entire process.

And to all of the educators who participated in pilot workshops, thank you for investing precious time in a new program and providing feedback about which activities worked, which ones did not, and where improvement was needed. Included in this group are the educators who met at Quiet Waters Park in Annapolis and at Horsehead Wetlands Center in Grasonville, as well as the group of educators from Fairfax County, Virginia. It is for you the educators that this book was created, to empower students to protect, preserve, and (when necessary) create wetlands.

KAREN L. RIPPLE

Contents

Part I: Background Information

Part II: Student Activities

INTRODUCTION

Natural wetlands are varied ecosystems that may be flooded for as little as 5% and as much as 100% of the growing season. Consequently, some wetlands will be wet every time they are visited, and others will usually be dry. These wetlands differ dramatically in their biota and their functions. Furthermore, there are many different types of wetlands that are between these two extremes. All of these wetlands offer educational potential as schoolyard habitats.

Most schoolyards, however, do not contain natural wetlands. Instead, they have soggy wet areas that are difficult to mow, or swales (often badly eroded) to direct runoff from the parking lot and playing fields, or maybe a series of pipes and culverts to divert roof runoff away from buildings. All of these situations have the potential of providing wetland habitat for creatures, a wetland refuge for native plant species, and an outdoor environmental study area for students. With information and guidance the negative aspects of a schoolyard can be turned into a positive learning experience.

Natural wetlands develop because of the presence of water during the growing season over long periods of time (hundreds, even thousands of years). Planned wetlands are very young wetlands that have been created by humans where no wetland existed. Initially, they may not be as effective as natural wetlands, but they do provide many wetland functions, and can be effective outdoor classrooms.

In addition to these created wetlands, planned wetlands include restored or enhanced natural wetlands. When some functions are deliberately restored to a damaged wetland, it becomes a planned wetland. A natural wetland in which some functions are enhanced (improved) also is considered a planned wetland.

POW!: The Planning of Wetlands was written for educators interested in creating, restoring, or enhancing, and then monitoring a wetland in their schoolyard. Educators in schools, refuges, nature centers, parks, and many other settings who are interested in designing and constructing planned wetlands will find *POW!: The Planning of Wetlands* to be a valuable reference guide as well as a source of hands-on activities that incorporate many job skills.

Part I of *POW!* deals with the technical aspects of planning wetlands. The first chapters provide background information about natural wetlands, wetland hydrology, and the different types of planned wetlands. Other chapters deal with permitting, design, construction, maintenance, and costs associated with planned wetlands. Another chapter provides useful information and drawings of native wetland plants.

Part II is a series of hands-on multi-disciplinary activities in lesson-plan format with reproducible student activity pages. (The binding allows the book to lie flat on a copy machine.) In using these activities educators will be able to guide students through the steps necessary to plan and monitor a wetland. The activities are suitable for grades five through twelve, with some adaptations for grades kindergarten through four. All activities are correlated with the National Science Education Standards (Appendix B).

Once the decision has been made to do a wetland project, it is important to determine the source of water for the planned wetland. The use of natural sources (such as runoff from the schoolyard) will require:

- considering several different possible sites for the wetland,
- modifying the topography of the land associated with the wetland site,

- calculating the quantity of water available over a given period of time, and
- determining which type of wetland can be sustained at the chosen site.

If an artificial water source (such as a faucet and length of hose) will be used, the above considerations are much less important.

Planning a wetland may seem a daunting task in light of the myriad responsibilities that educators tend to accumulate. Yet wetland projects do provide service learning and do qualify as student action projects. Since administrators and grant funding groups require assurance and some evidence of how the end product will appear, a pictorial review of one small planned wetland is included in Appendix C. To view additional wetlands planned by educators and students, visit Restoration Corner at www.wetland.org.

We hope that *POW!: The Planning of Wetlands* will instill respect for the complexity of wetlands, while encouraging efforts to restore these wondrous places. Based on our experiences, we firmly believe that students must be part of this process if they are to develop a sense of stewardship for wetlands--those marvelous places we now know to be finite and fragile. Enjoy the adventure!

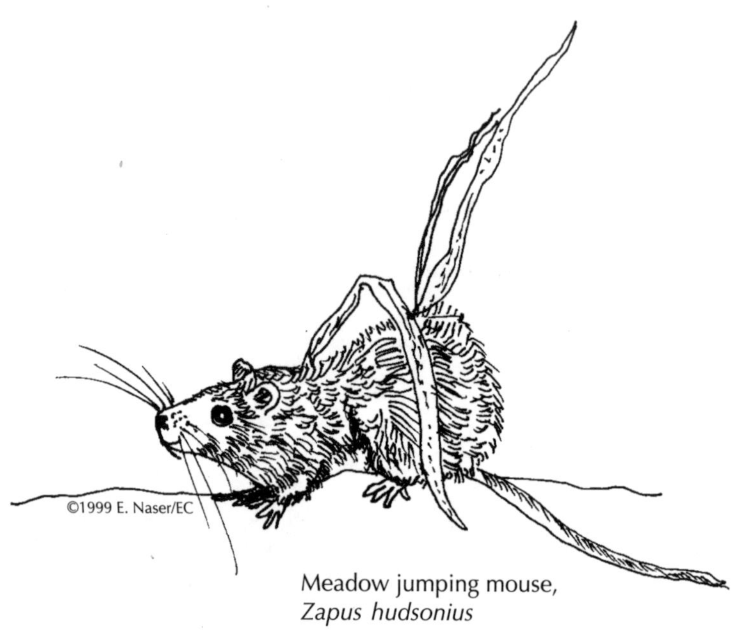

Meadow jumping mouse,
Zapus hudsonius

PART I

BACKGROUND INFORMATION

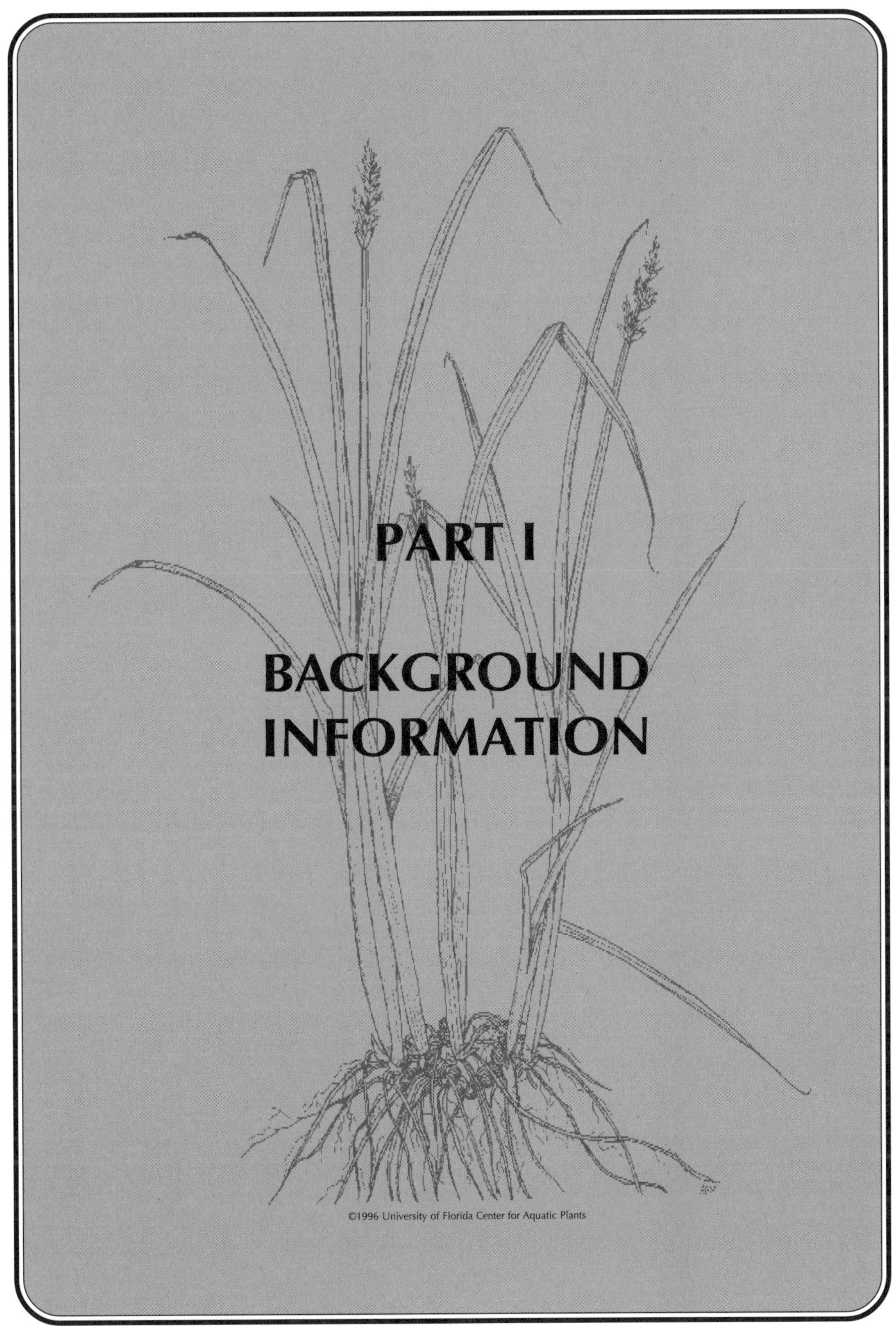

1 WETLAND BASICS

Wetlands! The very word conjures up visions of an alligator-infested swamp, or a mud hole smelling of rotten eggs. But wetlands are so much more than that. They are the kidneys and lungs of water bodies, removing suspended sediment and pollutants from the water and adding oxygen to the water. They are a nursery for juvenile finfish, a food source for shellfish, and habitat for thousands of organisms, including many endangered species. Wetlands stabilize sediment and reduce erosion along the banks of streams, rivers, lakes, and bays. Wetlands reduce flooding during storm events. Despite the aforementioned reputation, they add beauty to the land. Wetlands may only cover five percent of North America, but their contribution to the continent is enormous.

By definition of the U.S. Army Corps of Engineers (*Corps of Engineers Wetlands Delineation Manual* 1987), "wetlands are areas that are inundated or saturated by surface or ground water at a frequency and duration sufficient to support, and that under normal circumstances do support, a prevalence of vegetation typically adapted for life in saturated soil conditions. Wetlands generally include swamps, marshes, bogs, and similar areas." Many wetlands, however, are only seasonally or irregularly inundated and are likely to be dry when visited.

An area only need be inundated (flooded) or saturated for consecutive days equaling 7.5% of the growing season in order for it to be considered a wetland. The growing season varies in length depending on location. In general, the growing season is defined as the time period between the first bud break of vegetation and the first frost. Assuming the growing season in an area is 185 days, the site would only need to be saturated or inundated for approximately fourteen days in a row during the growing season to be considered a wetland. It could be dry for the remaining 171 days. Therefore, an area may be wet during the spring and dry during the summer and fall, yet still be considered a wetland. This example illustrates how profoundly the presence of water effects a habitat.

Wetlands have three special characteristics that, in conjunction with one another, make them different from other habitat types. These characteristics are the **presence of water** for a sufficient period of time (hydroperiod), to affect the **plant species** that become established, and to influence the development of unique **soil characteristics**. Therefore, wetlands are identified by their hydroperiod, vegetation, and soils.

VEGETATION

Wetland plants are often called aquatic plants and are classified as hydrophytes, which Webster's dictionary (1989) defines as perennial vascular aquatic plants having their overwintering buds under water, or plants growing in water or in soils too waterlogged for most plants to survive. Aquatic plants are able to grow in water, or in anoxic (oxygenless) conditions because, through a number of different physical and biochemical mechanisms, they are able to oxygenate their below-ground biomass to allow them to metabolize. Non-aquatic plants are unable to combat this lack of oxygen in the soil caused by water saturation, and will not survive when grown in water.

All plants, wetland and upland, are placed in five broad categories based on their water tolerance. The name or abbreviation of each category provides the wetland indicator status of each plant species, and are as follows:

Obligate upland (UPL). Fewer than 1% of these plants are found in wetlands. More than 99% of them occur in uplands. Plants such as cacti would fit into this category.

Facultative upland (FACU). From 1% to 33% of these plants are found in wetlands. Between 67% and 99% of them are found in uplands. Plants such as Kentucky bluegrass, eastern hemlock, and white oak are facultative upland plants.

Facultative (FAC). Plants with a similar likelihood of occurring in a wetland or an upland (33% to 67%). This category includes a wide variety of species like red maple, bayberry, box elder, and big bluestem.

Facultative wetland (FACW). From 67% to 99% of these plants are found in wetlands (1% to 33% in uplands). Examples of facultative wetland plants include river birch, jack-in-the-pulpit, and highbush blueberry.

Obligate wetland (OBL). More than 99% of these plants are found in wetlands. They are frequently the plants that we associate with wetlands. This group includes bald cypress, cattails, and water lilies.

The longer the hydroperiod in the wetland, the more likely that facultative wetland and obligate wetland plants will be present. In the earlier example, where water was only present for fourteen consecutive days, the most likely group to inhabit the wetland would be facultative plants.

SOILS

Wetland soils are known as hydric soils. These soils are different from other soils in that they have developed under anaerobic (low or no oxygen) conditions. Some wetland soils are dominated by organic material and are categorized as peats or mucks. These soils develop from the buildup of partially decomposed plant material, where a portion of the previous year's growth remains in place and becomes buried by a subsequent year's growth. Since decomposition (composting) of organic material is slower and sometimes nonexistent under anoxic conditions and in cooler environments, in many wetlands the organic material builds up much faster than it can be decomposed and transported from the wetland. Organic (peat) wetland soils generally develop rapidly in wetlands dominated by mosses (sphagnum or peat mosses) or herbaceous emergent vegetation (cattails, manna grass, cordgrass). To give some idea of how rapidly peat develops underground, realize that the annual below-ground biomass of many herbaceous emergent aquatics is equal to or greater than what is seen above-ground (the above-ground biomass).

Other wetland soils are dominated by mineral materials (sands, silts, and clays). These soils generally form in warmer climates, in wetlands dominated by woody vegetation (trees and shrubs), in wetlands which are only seasonally flooded or saturated (anoxic for only a portion of the growing season), and/or under conditions where organic matter is quickly exported from the wetland (e.g., fast-flowing streams, tidal action). Hydric soils which are mineral in nature tend to have two distinct characteristics: gleying and mottling.

Gleying is the process by which chemicals in the soil, such as iron and manganese, are reduced (oxygen is removed) in an anoxic environment, causing gray or black colors. This reduction process is very similar to processes that occur when making pottery. When pottery is made with an oxidized firing in the kiln, the finished pottery colors are usually bright. When the pottery is placed in a reduced (low oxygen) firing, dark, muted colors result.

Mottling appears as bright "spots" or concentrations in the soil. These mottles are frequently reddish and look rusty. They are composed of iron oxide, which is rust. Why would oxidized iron be found in a wetland that is anoxic? Remember, most wetlands are not inundated or saturated for the entire growing season. Many wetlands are only wet in the spring, then dry out as the summer progresses. During dry periods, iron compounds in the wetland soils are oxidized to iron oxide. In the very same process that forms rust on garden tools and wrought iron, the iron in the soil rusts. This rust then settles out in small spaces or bands in the soil column, transported by a fluctuating water table, or downward movement of water through the soil from the surface. This is known as iron segregation or mineral segregation. This condition is found commonly in wetlands that are only seasonally wet.

Both gleying and mottling are produced by very stable chemical reactions. Therefore, after a soil is mottled or gleyed, even if conditions change, the gleying and/or mottling will remain a characteristic of the soil. If a seasonally flooded wetland is drained by ditching so that water will never return to saturate its soils, the mottled and/or gleyed conditions will persist. Consequently, one cannot claim that gleying or mottling in soils is evidence that wetland hydrology presently exists.

HYDROLOGY

A wetland's hydrology is its hydroperiod (the duration of flooding or soil saturation) and its depth of flooding. Since a wetland may consist of more than one elevation, there may be more than one hydroperiod and depth of flooding. Hydrology determines the wetland's vegetation and benthic communities. It is the hydrology that determines what plant species will become established at the site, what wildlife will utilize the site, and what invertebrates will inhabit the site's soils. Hydrology creates the environmental conditions necessary for the development of wetland (hydric) soils.

There are many different types of wetlands, each with its unique hydrology. Wetland hydroperiods range from permanent flooding (inundation or soil saturation), to regular flooding twice daily by the ocean tides, to about a week of flooding during the entire growing season. Consequently, there are wetlands that are unlikely to be wet when visited and others which always will be wet.

All wetlands fall into one of the following four hydroperiods or periods of flooding (inundation or soil saturation):

Semipermanent to permanent. Flooded from 76% to 100% of the growing season.

Regular. Flooded from 26% to 75% of the growing season.

Seasonal. Flooded from 13% to 25% of the growing season.

Irregular. Flooded from 5% to 12% of the growing season (9 to 22 days in a 185-day growing season).

Although it was stated earlier that in order for a site to be definitely considered as having wetland hydrology it must be saturated for consecutive days equaling at least 7.5% of the growing season, under some circumstances wetlands may form when flooded 5% of the growing season. Not all sites that are flooded for this length of time will be wetlands, but they might be. Under these conditions both hydric soil characteristics and aquatic vegetation must be present for the site to be considered a wetland.

Irregularly flooded wetlands are frequently the most confusing ones. During most of the year the site is dry and contains facultative plant species. People visiting these wetlands may come to the conclusion that they are uplands. However, hydric soil characteristics will be present.

Seasonally flooded wetlands are flooded 23 to 46 consecutive days in a 185-day growing season. The hydroperiod supports facultative to obligate wetland plants. These wetlands generally are recognized as such by most people. Both irregularly and seasonally flooded wetlands often contain drought-tolerant plants.

Regularly flooded wetlands are flooded 47 to 139 consecutive days in a 185-day growing season. These wetlands are dominated by facultative wetland and obligate wetland plants. Wetlands having these hydroperiods are often associated with water sources having an abundant water supply; e.g., oceans, lakes, rivers, perennial streams, and often groundwater.

Semipermanently to permanently flooded wetlands are flooded 140 or more consecutive days in a 185-day growing season. Although these wetlands are flooded longer than the regularly flooded wetlands, they are very similar to them in other respects.

PLANNED WETLANDS

A planned wetland is a constructed, restored, or enhanced wetland.

A constructed wetland is a wetland that is created on a non-wetland site. The wetland construction site may be an upland area that is physically altered to receive water from a natural or an artificial source, or it may be an open water area that is filled, usually by dredged materials, and graded to wetland elevations.

A restored wetland is a wetland that has been returned from an impacted or destroyed condition to a less impacted and fully functional condition. Generally, wetland restoration includes restoring both the wetland hydrology (hydroperiods and depths of water) and the wetland vegetation communities.

An enhanced wetland is a wetland whose functions are altered through physical changes to yield an overall improved wetland or a wetland whose enhanced function is needed in the region. For example, a high elevation wetland in a severely polluted waterway does not have the needed ability to improve water quality due to its short hydroperiod. By reducing its elevation and increasing its hydroperiod, this needed function is enhanced; however other functions, such as wildlife habitat, are diminished. In a second example a wetland that is choked by invasive and undesirable plant species such as purple loosestrife (*Lythrum salicaria*) and common reed (*Phragmites australis*) can have its biodiversity enhanced by removing this vegetation and replacing it with more diverse native species.

There are many benefits to having a planned wetland on a schoolyard. It can provide wildlife habitat, water quality improvements, waterway (e.g., stream) bank erosion control, flood control, and other benefits. It will be a living classroom for the school and a source of pride for those who participate in its design and construction. It can provide opportunities for research in botany, biology, water chemistry, zoology, and other disciplines. The uses of a planned wetland for learning are only limited by one's imagination.

A wetland may exist on some schoolyards and provide an opportunity for enhancement and even restoration, if there is evidence of prior disturbance. However, it is unlikely that there will be evidence that a wetland once existed but has been destroyed and the area put to other uses, such as a building site or athletic field. The greatest opportunity for a planned wetland on a schoolyard will be one of constructing a wetland. Because of the logistics of use and reliability, in many instances the source of water will be artificial.

WATER SOURCES

The water that supports a wetland can come from several sources, often in combination. These water sources include:

Ocean. The oceans of the world are the salt and brackish water sources for all coastal and estuarine tidal wetlands. These sources can be considered to provide an unlimited supply of water to tidal wetlands. Ocean waters are even responsible for the tides in tidal freshwater wetlands associated with rivers whose mouths are along the coast.

Surface Water. The source for the water in ponds, lakes, streams, and tidal and nontidal freshwater portions of rivers is surface water (stormwater runoff) or a combination of surface water and groundwater. These bodies of water may be considered to provide an abundant supply of water to wetlands, and a water budget generally need not be calculated for the planned wetland to ensure that the desired hydroperiod will be achieved. Groundwater and surface water would be the sources of water for the many types of wetlands associated with these bodies of water.

Many wetlands are supported only by surface water. These wetlands do not have a direct connection with the water table, and rely solely on rainwater and snow melt for their water supply. The supply of this water may be considered potentially limited and a water budget should be calculated for the planned wetland to ensure that the desired hydroperiod can be achieved. These wetlands usually form in depressional areas, and hold water because the underlying soils are impermeable, consisting largely of silts and clays, or because other barriers to infiltration (e.g., bedrock) lie below the surface. These wetlands commonly are referred to as perched wetlands. The hydroperiods of these wetlands vary from permanently to irregularly inundated depending upon their size, the size of their watersheds, and the frequency and quantity of precipitation.

Groundwater. There are three conditions where groundwater is the primary source of water for wetlands.

The first condition is a groundwater spring, which might come from an artesian aquifer that discharges water to the ground surface. Depending upon the topography of the land, this discharged water may flow down a gradient to pond in depressions having impermeable soils that contain perched wetlands. Or it may flow down a gradient, supporting wetlands until its infiltration into the soil reaches a depth where wetlands no longer can be supported. In either instance, the groundwater supports wetlands having hydroperiods that relate to whether the spring is permanent or seasonal. If the spring is seasonal, there may be a surface water contribution to the water supporting the wetland. In other instances, water from the spring may discharge directly or flow directly into a stream and its associated wetlands (**Figure 1.1**).

The second condition is where the water table meets a depressional area formed of permeable unconfined soils. When the water table rises and intercepts the depressional area, water is discharged into the depression and the water level rises to the height of the water table. Whether the depressional area permanently or seasonally intercepts the water table will dictate the wetland's hydroperiod and its biotic communities (**Figure 1.1**). If surface water enters the wetland when it is dry, the water will infiltrate to the water table. If there is groundwater in the wetland when surface water enters, the water level will rise temporarily and then lower through infiltration to the water table level.

The third condition is where the water table does not usually reach the surface of the ground, but lies just beneath its surface during part of the growing season. Under these conditions, wetland plants will be supported and hydric soils will develop near the surface. Often, these groundwater-supported wetlands will have their soils saturated during the wettest part of the growing season and will dry out during the balance of the season. The height of the water table in these instances is dependent on the amount of precipitation and how much of it infiltrates through the soil to the water table.

The supply of groundwater in these three situations may be either abundant or potentially limited, depending upon:

- the size of the planned wetland to be supplied,
- the size of the applicable ground watershed,
- the volume of water contained in this watershed,

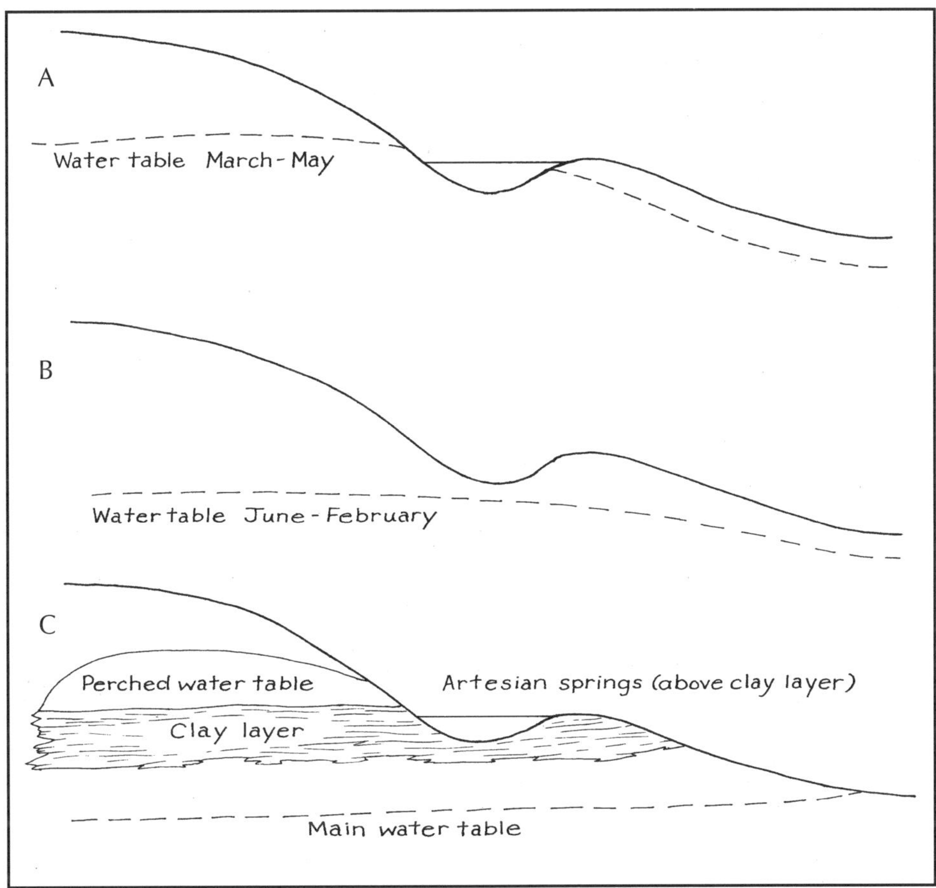

Figure 1.1 Illustrated are two depressional wetlands. One is a seasonally flooded wetland where the water table rises and intercepts the depression (A) during the high water months (March-May), but is dry (B) during the low water months (June-February). The other is a perched wetland (C) supported by a groundwater spring.

- the rate of groundwater discharge into the planned wetland, and
- the rate of groundwater recharge.

Other than determining the size of the wetland, there are currently no reliable techniques for measuring these inputs. Consequently, it is very difficult to calculate a water budget for a wetland supported solely by groundwater.

Direct Precipitation. Depressional areas that have no watershed may support wetlands if their soils are impermeable, like the perched wetlands described above. If they do, they are supported by direct precipitation and are likely to be irregularly inundated at best. Their vegetation and benthic communities would have to be drought-tolerant. Direct precipitation may contribute to the water supply for most wetlands; however, it generally is a minor contributor and only in the condition cited above is it the sole contributor.

Well or Local Water Supply. Schoolyards may not have the natural water supply to support wetlands of the size, location, and type that the students are interested in planning and studying. Additionally, students may want to study aspects associated with more than one wetland type concurrently or in different time intervals. In order to construct a desired planned wetland in a schoolyard, it may be necessary or practical to install a well for the water supply or to use the local dechlorinated water supply in order to obtain the desired hydrology.

COMMON WETLAND TYPES

Swamps, Bottomland Forests, and Riparian Forests

Swamp seems to be the universal name given to wetlands by the general public. But a swamp is actually a type of wetland dominated by trees and shrubs. Swamps have a variety of common names across the North American continent including bottomland forest, lowland forest, loblolly forest, and wet woods. Regardless of the colloquial name, they all have one thing in common; they are wetlands dominated by woody vegetation. They are forested wetlands.

In southern swamps or bottomland forests, wildlife abounds (**Figure 1.2**). Almost 300 bird, 80 reptile and amphibian, and 50 mammal species use these forests for resting, feeding, and raising their young. In addition to cottonmouths and alligators, these swamps support a variety of endangered and threatened species such as the wood stork, Bachman's warbler, and the bald eagle. Trees in these southern bottomland forests range from the more common species, such as red maples, box elders, and sweet and black gums, to the truly southern species, such as the bald cypress.

Further north and west along the North American continent, forested wetlands associated with rivers and streams tend to become narrower in width, and more closely associated with a defined river or stream channel. These wetlands are generally called riparian forests (**Figure 1.3**). These forests provide important wildlife habitat, migration corridors, and breeding and feeding areas. These wetlands are dominated by a variety of species, including black willow, red maple, and sycamore. In the West, cottonwoods are an important component of the riparian forest. In the more northern climes, ice scouring can be a significant natural impact to riparian wetlands. Species like black willow and red maple have adapted to this by being able to readily sprout from stumps and downed trees damaged by ice flow. Swamps are found not only near rivers and streams, but lake shores also frequently support swamps. These swamps may be in the shallow water along the shore, or just out of the lake but supported by groundwater that is influenced by lake levels.

Inland, away from the rivers, streams, and lakes, the landscape is dotted with depressions that are dominated by trees and shrubs. These forested wetlands can be from a few square meters to hundreds of hectares in size. They may be supported by groundwater, like the Pitch Pine Lowlands of New Jersey, or by surface water runoff, like the Delmarva Bays of Delaware. Many of these wetlands have short hydroperiods and are dominated by facultative plant species. These are the wetlands that are very frequently impacted by development. They are dry during the summer months and often wet only in the spring. Many people do not recognize them as wetlands because of the lack of water in the summer and the dominance of facultative trees and shrubs. Regardless of the time of year, the soils in these swamps will be hydric, evidence that the site is a wetland, provided there are no signs of hydrologic manipulation. These

Figure 1.2 In a bald cypress swamp trees have buttressed bases. Buttressing increases stability in the soft saturated sediments by making the trees less top heavy. A diversity of waterfowl feed on the duckweed that floats in the water.

Figure 1.3 In a Northeastern riparian forest red and silver maples, American elms, and sycamores are among the many trees adapted to seasonal flooding. Sometimes the trees are zoned according to their tolerance to different flooding frequencies and hydroperiods.

©2000 C. Barcomb/EC

Figure 1.4(a) A small pond in Maryland that has a shallow shoreline (fringe) marsh with varying hydroperiods (irregularly flooded). Pictured in the foreground is pickerelweed, *Pontederia cordata*.

Figure 1.4(b) The same small pond in Maryland has a deep emergent marsh shelf that is permanently flooded. Pictured is lizard's tail, *Sauruus cernuus*.

inland swamps can be very important wildlife habitat, and support a host of threatened and endangered species including the Pine Barrens tree frog, the bog turtle, and swamp pink.

Not all swamps are dominated by trees or by a mixture of trees and shrubs. Many swamps are dominated by shrubs, and are called scrub-shrub wetlands. These wetlands provide dense cover for wildlife and are important for soil stabilization. Shrubs that dominate these wetlands include blueberries, cranberries, and rhododendrons. These wetlands can have any water source other than ocean water.

Nontidal Freshwater Marshes

A marsh is distinguished from a swamp by the fact that the dominant vegetation is herbaceous instead of woody, and is emergent (coming out of the water) or floating. Submerged aquatic vegetation (SAV) grows underwater and generally is considered to be a wetland type that is separate from a marsh. Marshes are the most common wetland type in North America and are just as varied in their water sources and hydroperiods as are swamps.

Marsh vegetation can be broadly divided into two categories: persistent emergent and nonpersistent emergent. Persistent emergent vegetation remains erect in the marsh even after the plants have gone dormant for the winter. Cattail is an example of a persistent emergent species. Cattails can be seen standing throughout the dormant season until they are replaced by new plants during the next growing season. Examples of nonpersistent emergent species are duck potato, arrow arum, and pickerelweed [**Figures 1.4(a) and 1.4(b)**]. The above-ground parts of these species decompose rapidly after the plant goes into dormancy, and there is no evidence of their presence during the dormant season. The perennial parts (rhizomes, bulbs, tubers) of the persistent and nonpersistent species remain underground during the dormant season. Nonpersistent emergent species are more common than persistent emergent ones.

Marshes are associated with all nontidal bodies of water. They have hydroperiods that range from permanently flooded to irregularly flooded (**Figure 1.5**). In the Northeast, those irregularly flooded marshes tend to convert to forested wetlands or swamps over a period of time. Marshes

Figure 1.5 A stream and associated marsh near Conway, Massachusetts.

Figure 1.6 A tidal salt marsh along the mid-Atlantic coast.

©2000 C. Barcomb/EC

along roadside ditches, although generally irregularly flooded, tend to remain herbaceous because of periodic mowing.

Marshes need at least six hours of direct sunlight daily during the growing season in order to maintain normal plant morphologies and densities. As the duration of direct sunlight is reduced, plants elongate and their densities thin, eventually leading to no vegetation at two to three hours of direct sunlight daily. Consequently, marshes rarely are found in swamps, bottomland forests, and riparian forests, unless large open areas develop due to tree blowdown, etc.

Invasive and undesirable species are more of a problem in freshwater marshes than in swamps. As the desirable species in a marsh become displaced by invasive ones, the functional value of the wetland diminishes. This is because the drop in species diversity of the marsh diminishes its wildlife function as well. Examples of invasive species include common reed, purple loosestrife, reed canary grass, and cattail.

Tidal Wetlands

Tidal wetlands are supported by tidal water (**Figure 1.6**). The gravitational pull of the sun and moon on the earth's oceans causes a cyclic water movement, yielding twice daily low and high tides. The area between the low tide and the high tide elevations is known as the intertidal zone. The area between the usual high tide and the spring tide elevations (when the moon and sun are aligned, the tide springs to unusually high elevations) is called the supratidal zone. It is in these two zones that the majority of tidal marshes and tidal swamps are found.

Most of us think of tidal wetlands as salt marshes (tidal wetlands flooded by salt or brackish water of the coasts and estuaries throughout North America). However, there are tidal freshwater marshes and swamps as well. All rivers that have their mouths at the coast or in tidal bays have tidal freshwater at various distances upriver. As the high tide comes in, the coast/bay water pushes up into the rivers and changes the direction of the river's freshwater flow. As the freshwater is pushed back upriver, the water elevation rises, causing a high tide of freshwater. As the coast/bay water recedes to a low tide, the river again reverses its flow and its water level drops, causing a freshwater low tide.

Along the coasts of the Atlantic Ocean, the Pacific Ocean, and the Gulf of Mexico are vast areas of wetlands. The intertidal and supratidal zones of these wetlands support mostly salt marsh cordgrass (*Spartina alterniflora*) and salt marsh hay (*Spartina patens*), respectively. On the West Coast of the United States, the intertidal zone is dominated by native Pacific cordgrass (*Spartina foliosa*). On the East Coast of the United States, the supratidal zone of these saltwater and brackish water wetlands sometimes is occupied by the tidal scrub-shrub wetland plant commonly called high tide bush. High tide bush refers collectively to two similar looking shrubs known as groundsel tree (*Baccharis halimifolia*) and marsh elder (*Iva frutescens*). On both coasts of Southern Florida, on the southern coast of Texas, and in Hawaii the upper intertidal and the supratidal zones are often occupied by scrub-shrub wetlands known as mangrove swamps or mangrove forests (**Figure 1.7**).

Bog turtle,
Clemmys muhlenbergi

Figure 1.7 A drawing of a red mangrove forest showing its arching prop roots.

Figure 1.8 A tidal freshwater marsh on the Delaware River near Philadelphia, Pennsylvania.

Upriver, where the tidal water changes from brackish to freshwater, are tidal freshwater marshes and tidal freshwater riparian forests or swamps. The vegetation in these wetlands is similar to that in the nearby nontidal wetlands (**Figure 1.8**).

Tidal and nontidal marshes develop peat banks over time that extend above the original ground surface. This is because these plants produce as much below-ground as above-ground biomass each growing season. There is more resistance to this below-ground peat production forcing its way further downward into consolidated sediments than to its moving upward. Consequently, a peat bank is formed that can be seen at low tide in tidal marshes and during water drawdowns in nontidal marshes.

The tidal wetland hydroperiods may be affected as the peat banks develop, provided the rate of sea level rise is different than the rate of peat bank development (**Figure 1.9**). In nontidal wetlands, the hydroperiods and/or the depths of water will always be reduced as the peat bank develops.

Bogs and Fens

Bogs and fens are special and complex types of wetlands that are mostly located in the cool, moist, and glaciated regions of the United States and Canada. They are also known as peatlands. Bogs and fens slowly accumulate peat from non-decaying above-ground plant production and from below-ground plant production. Peat usually accumulates under conditions where plant biomass decomposition is largely eliminated by anoxic con-

Figure 1.9 A twelve-inch thick peat bank developed in this planned salt marsh cordgrass zone during its first twenty years.

ditions, low temperatures, and low pH (acidic conditions). Bogs generally contain carpets of sphagnum moss (which releases hydrogen ions that lower pH) from which grow cranberries, orchids, pitcher plants, and many other plants. Fens are less acidic and more fertile than bogs. In fens sedges largely replace sphagnum as producers of peat. Because of their complex character and their slow production of peat, bogs and fens are unlikely candidates for schoolyard wetland habitats. Consequently, the planning of bogs and fens is not within the purview of this book.

WETLAND FUNCTIONS AND VALUES

Wetland functions and values are often different. A function is a task or work performed by a wetland, regardless of how society values that function. A wetland's function is based on its structure or composition. For example, if a wetland is dominated by emergent vegetation mixed in an approximately 50:50 ratio with open water, it is good habitat for waterfowl. This is because the vegetation provides cover for the waterfowl and the vegetation/water interspersion provides access to the water for feeding and resting. The structure of the wetland creates the conditions in the wetland so that it can function as good waterfowl habitat.

A value is the worth that humans place on a particular function of a wetland. Societal values may change over time. For example, some prehistoric humans placed a value on bogs as a place to sacrifice other humans to curry favor with their gods. Modern society disdains the practice of human sacrifice and no longer places that value on wetlands. Today, society values those very same wetlands for preserving the ancient sacrificial remains for modern scientific inquiry. A wetland that functions as waterfowl habitat will be valued by some as an area for active (hunting) and passive (bird watching) recreation. An economic value could also be assigned to waterfowl habitats if they are rented out for hunting. Consequently, a wetland function often can be considered a wetland value as well.

Humans assign value to many wetland functions. For example, wetlands covered completely by dense, persistent emergent vegetation and which are in contact with the water column, are effective at improving water quality. This is because the dense vegetation slows down the water flowing through it, promoting sedimentation (the settlement of particles suspended in the water). During the growing season the dense vegetation will also uptake nutrients that may be present in the water column. This structure or composition of the wetland allows it to function to improve water quality. Good water quality is highly valued by humans. Therefore, water quality improvement is both a function and a value of wetlands.

A wetland cannot perform all functions effectively and simultaneously. It was noted that good waterfowl habitat included a 50:50 ratio of vegetation to water interspersion. A wetland with a high capacity for improving water quality needs dense, persistent emergent vegetation covering most of it. Consequently, a wetland that is structured best for water quality improvement will not function as an optimal waterfowl habitat. Following are the generally recognized functions of wetlands.

Shoreline Bank Erosion Control

Wetlands often occupy the interface between upland banks and open water. In coastal situations, as well as along lake, pond, river and stream banks, wetlands buffer the upland banks from the erosive forces of water. These forces may come from waves or currents interacting with the bank face. Under these conditions, the wetland plants absorb the wave or current energy, thereby reducing the energy interacting with the bank face. This leads to a reduction in the erosion rate and to an increase in erosion control. This is also a water quality function, since a reduction in erosion also means a reduction in sediments entering the water body. The value of this function is that it both protects water quality, and preserves property values by ensuring that waterfront property is not lost to erosion (**Figure 1.10**).

Sediment Stabilization

Wetland plants stabilize the substrate that they are growing in, thereby reducing its erosion. This function is maximized by dense above-ground and below-ground biomass. The above-ground biomass slows the water velocity and the below-ground biomass protects the substrate by its root mat. Settlement of sediment particles in the water column is also promoted through the reduction of water velocity by the wetland. Sediment stabilization reduces land erosion and protects water quality, both of which are valued functions.

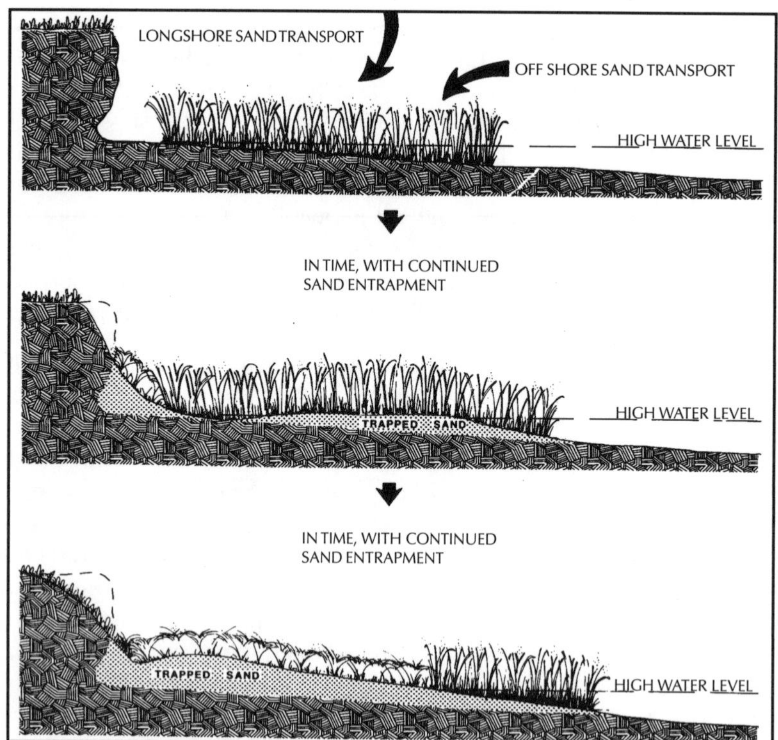

Figure 1.10 The wetland reduces the energy of waves and currents interacting with the bank face, thereby reducing bank erosion rates and depositing suspended sands. As suspended sands are deposited, the shore increases in elevation. This in turn decreases the frequency with which waves and currents interact with the bank face, thereby decreasing the rate of bank erosion.

Water Quality Enhancement

The surface water benefits of this function have been discussed above, but wetlands are also important for groundwater quality protection. Wetlands usually occupy the low spot on the topography. Surface water flows to these wetlands. If the bottom sediments in the wetlands are sandy, there is a good possibility that the water entering the wetland will leave the system through infiltration into the water table. As the surface water passes through the wetland, suspended particles settle out. Plants absorb dissolved nutrients, and a host of fungi and bacteria go to work on dissolved impurities in the water as it filters through the wetland soils. These soils act as a chemical filter by adsorbing dissolved impurities and as a physical filter by removing particulates. By the time the water reaches the water table, it is a lot cleaner than when it entered the wetland. This function, as it relates to groundwater purity, also has value for well-water users.

Flood Water Storage and Flood Flow Alteration

Many wetlands can store or detain storm waters and thereby reduce the frequency and duration of flooding of neighboring lands. Riparian wetlands and dense marshes obstruct and slow stormwater flow, and greatly reduce the heights and extend the time of downstream flood water peaks.

By retaining, detaining, and slowing flood waters, wetlands can release these waters over extended periods. These flood waters augment base flows of streams and rivers, thereby protecting the downstream aquatic life by reducing the likelihood and severity of dry spells. This function is of great value to all those living near floodplains and areas prone to flooding (**Figure 1.11**).

Groundwater Recharge

A wetland purifies the water that it contributes to the water table. Wetlands recharge the groundwater, replacing water that was removed for human uses, including potable water and crop irrigation.

Groundwater Discharge

When the water table is above the bottom of a wetland which has a permeable substrate, groundwater discharges into the wetland. The discharge function provides the necessary water to support the wetland. Some wetlands are groundwater discharge wetlands during the wet season, and groundwater recharge wetlands during the dry season.

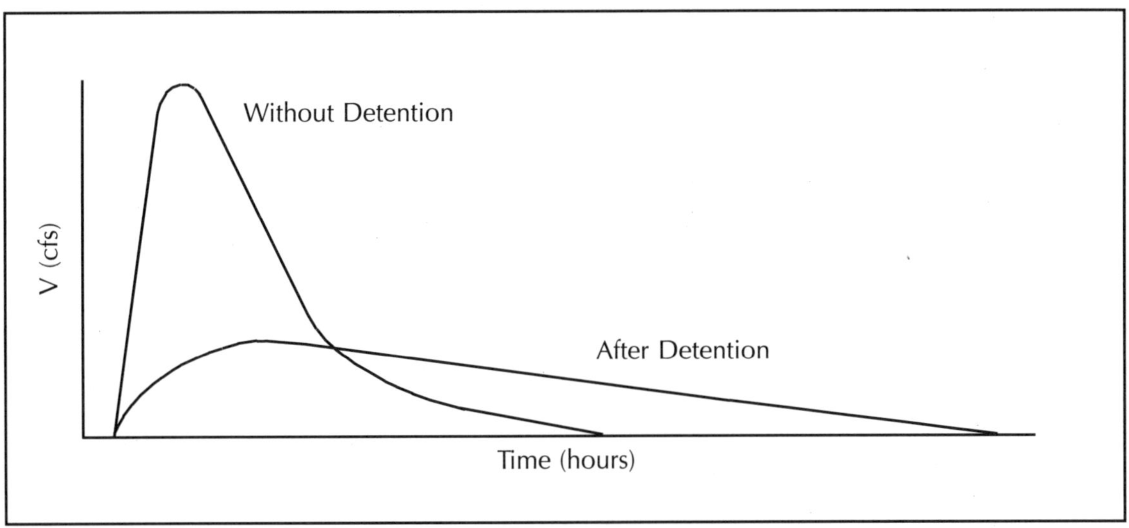

Figure 1.11 A wetland decreases flooding downstream through flood flow alteration, detaining flood waters and reducing the peak velocity (V) of runoff.

Wildlife Habitat

The importance of wetlands for waterfowl has previously been discussed, but a multitude of other wildlife also depends on wetlands. From moose to snails, wetlands are important habitats.

It is the structure or composition of the wetland that determines how well it will function as wildlife habitat and what specific wildlife will utilize it. A marsh that is permanently flooded will function well to improve water quality, but it does not function well as wildlife habitat. Diversity is needed in the structure of the wetland as well as in its hydroperiod to maximize its functional potential as wildlife habitat. A wetland that is partially dry for much of the year and that offers groundcover as well as trees and shrubs will provide cover, nesting potential, isolation from predators, and food for wildlife. Such a wetland could have a high wildlife habitat function.

Biological Diversity

Not only do many common animals inhabit wetlands, but fully two-thirds of all endangered species depend on wetlands for their survival. For many reasons humans now place a premium on the protection of endangered animal and plant species, and with them, their genetic heritage.

Endangered species protection can best be accomplished by protecting existing critical habitat from further damage or destruction. Enhancement and restoration of damaged habitat, however, also have value, particularly in linking undamaged areas to create wildlife corridors. Even creating habitat where none existed has a role in returning endangered species to viable population levels.

Fisheries Production

Close to three-quarters of all commercially and recreationally important fin and shellfish are directly dependent on wetlands at some point in their life cycles. Wetlands provide breeding, feeding, spawning, and nursery habitat for these species. For a wetland to have a high potential to provide the fisheries production function, it must have extended hydroperiods, barriers to protect juveniles from adult fish, and structure for eggs to rest on. Both tidal and permanently flooded non-tidal marshes have high potential to provide this function.

It has been estimated that $24,000,000,000 was spent in the U.S. for recreational fishing activities in 1991 alone (last year for which data is currently available). This does not include the commercial fishing figures. Wetlands provide both economic and recreational values.

In addition to the values associated with wetland functions, society places the following additional values on wetlands:

Food Production

Rice, a wetland plant grown in farmed wetlands, is the most important grain crop in the world. Such favorites as cranberries, water chestnuts, and the fish production discussed above, as well as aquaculture, make wetlands an important source of food worldwide.

Aesthetics

Many people find wetlands beautiful and peaceful. They enjoy relaxing in and around them. They enjoy watching the wildlife that are attracted to them. Wetlands are considered by many to add monetary value to their property because of their aesthetics.

Recreation

Wetland-related recreation contributes substantially to many local economies. Fishing, hunting, and trapping are popular activities. Canoeing, hiking, bird watching, and nature photography also have many proponents.

Open Space Buffers

Wetlands often supply much needed open space, particularly in an urban or suburban setting. They give people living near them "breathing space," help relieve the congestion of urban living, and serve as noise buffers.

Places for Scientific Inquiry

Wetlands are used as living laboratories by individuals of all age groups and academic backgrounds. From the kindergarten class catching tadpoles along the wetland edge, to long-term university studies, wetlands provide unique educational experiences and opportunities.

DYNAMICS OF WETLANDS

There is a classic ecology textbook example of succession. It begins with a pond. The pond slowly fills in and becomes an emergent marsh, the marsh slowly builds up peat and fills in even more. Then trees develop on the site, and eventually, the pond becomes upland. This is called directional or traditional succession. We now know that this type of succession has limited application in wetlands and that marshes are not slowly but surely moving towards becoming upland. Wetlands may change over time, but they usually remain wetlands. Wetlands are generally dynamic ecosystems, places that are constantly changing. These changes are often cyclical or pulsed and caused by natural or human-induced disturbances.

Some of the cyclical or pulsed causes of the dynamics of or changes in wetlands are given below.

Animal Eatout

There are many examples of animals contributing to changes in wetlands by their feeding habits. **Figure 1.12** shows a ten-acre, monotypic, arrow arum (*Peltandra virginica*), tidal freshwater marsh that twelve years earlier was a monotypic, narrow-leaved cattail (*Typha angustifolia*), tidal freshwater marsh. Feeding on the cattails by muskrats (*Ondatra zibethicus*) led to the conversion of the peat substrate of the cattails to an organic soup which cattails cannot recolonize because of its anoxic condition. However, arrow arum toler-

Figure 1.12 A monotypic arrow arum tidal freshwater marsh that formed after a muskrat eatout of a monotypic cattail marsh.

Figure 1.13 A uniformly vegetated two-acre one-year-old marsh.

Figure 1.14 The same salt marsh after several hours of feeding at high tide by a flock of Canada geese.

ates this anoxic condition and is not eaten by animals because of the oxalic acid content in its tissue. Consequently, arrow arum recolonized the organic soup.

Figure 1.13 shows a one-year-old two-acre salt marsh that is uniformly vegetated by cordgrass. **Figures 1.14 and 1.15** show this same marsh after about four hours of feeding by a flock of Canada geese (*Branta canadensis*).

Canada geese do not fly into wetlands that are uniformly vegetated with herbaceous plants that are less than one-foot tall or with woody vegetation variably spaced about twelve feet apart, for fear of not landing safely.

Geese will fly into a mowed area or into open water, and then walk or swim into the wetland to feed. If the wetland is protected by a nylon line exclosure fence as shown in **Figure 1.16**, wild Canada geese will not swim or climb through the fence. Domesticated geese will, however, get through the fence and feed on the vegetation.

As a marsh matures with an underlying rhizome (peat) rug, the percentage of live perennial parts in the peat decreases. Canada geese will then

Figure 1.15 Another view of the marsh on the previous page following the Canada geese eatout. Note the swales (depressions) caused by geese washing away sediment at high tide to feed on the plants' rhizomes.

Figure 1.16 This is an effective Canada goose exclosure fence. The nylon lines are spaced six inches apart and placed from low to high water elevations in the wetland.

concentrate their feeding on the edges of the marsh, because washing away sediment to reach live perennial plant parts becomes more difficult and less rewarding.

Other animals that can severely affect wetlands and contribute to their dynamics are snow geese (*Chen caerulescens) [marshes]*, deer (*Odocoileus virginianus*) [forested wetlands], nutria (*Myocastor coypus)* [marshes], and beavers (*Castor canadensis*) [stream corridors].

Unstable Peat Production

Figure 1.9 shows the peat buildup in a twenty-year-old constructed salt marsh. This salt marsh is not subject to significant wave energy, and the developed peat bank was found to be stable. Depending upon sediment compaction, peat banks may rise above the sediment surface after two to four years of vegetation of the sediment by rhizome-propagating emergent plants. During the period that the peat remains underground, the marsh on shallow, sloping shores may be stable to fetches (open water distances) of up to fifteen miles.

Once the vertical peat bank rises above the sediment surface, however, the marsh's days are numbered if the fetch is over one mile. **Figure 1.17(a)** shows the beginning of the loss of about five acres of salt marsh after a peat bank rose one inch above the sediment surface. This salt marsh was stable for four years with a winter wind fetch of over five miles. However, once its vertical peat bank emerged from the sediment surface, much of the wetland washed away as an organic rug during the wind event shown in **Figure 1.17(b).**

Even with a fetch of less than one mile, marsh peat banks are becoming increasingly vulnerable to erosion. This is because there is an increasing use of sheltered waters for recreational purposes. Presently speed boats and jet skis are the prime culprits.

Plant Disease and Insect Problems

The gypsy moth caterpillar can be particularly damaging to swamps that contain an abundance of oaks. Low gypsy moth populations may reduce the species diversity of a swamp by selectively killing the oak species. However, in high populations, gypsy moth caterpillars will damage most woody species.

Wetlands are vulnerable to being affected by disease, particularly wetlands having a low diversity of plant species. The intertidal zones of salt marshes, which often are monotypically vegetated by cordgrass, are prime candidates for being severely affected by disease. One disease that has been known for decades to infect cordgrass (*Spartina alterniflora*) and common threesquare (*Scirpus pungens*) is a rust fungus yet to be genetically identified. If noticed in time, this rust can be controlled by most commercial fungicides that are labeled to control rust. Rust may affect these species during any time of year. However, spring and fall are the most likely times because temperatures and humidity levels in marshes are high then.

Figure 1.17(a) A young stable marsh before its peat bank emerges.

Figure 1.17(b) A four-year-old salt marsh that is being washed away after its peat bank emerged one inch above the sediment surface.

The rust spores are red to orange in color and easily identified on an infected plant. When the spores die, their color changes to black permanently (see **Figure 1.18**). Consequently, one can look at the plant litter that has washed out of a marsh and determine if any of the plants were rust-infested and to what extent.

The worst time for a marsh to be seriously affected by rust is in the spring before the plants have produced perennial parts (new shoots and rhizomes). In these instances, the plants can be killed with no hope for regeneration. Under severe infestation, the marsh could be lost.

If a marsh has been infected by rust in the fall, the below-ground perennial parts (shoots and rhizomes) would be mature and isolated from the spores. Just the above-ground stems and leaves would become infected and have their deaths accelerated thereby. Since the perennial plant parts would not be affected, the marsh would resume growth the following spring.

Marshes have been lost in Maryland, Delaware, and New Hampshire due to spring infestation by rust. A salt marsh in Delaware was 99% lost, but became 85% restored within three years (personal observation). Several infected marshes in Maryland never recovered. However, salt marshes from Maine to Virginia that encountered serious autumn rust infestations showed no effects the following spring.

Figure 1.18 Dead rust spores on the leaves and stems of cordgrass

An illustrative case involves a cordgrass marsh with an island of common three square that was constructed by Environmental Concern in Baltimore, Maryland in 1981 (**Figure 1.19**). In August 1989 the marsh appeared totally dead (**Figure 1.20**). By May 1990, when all surrounding vegetation had fully leafed out, there was still no sign of plant life in the marsh (**Figure 1.21**). However, by August 1991, the marsh had totally recovered, with the common three square island in place surrounded by cordgrass (**Figure 1.22**).

The surrounding plant litter showed definite signs of rust infestation (streaks of dead black rust spores on the leaves and stems), suggesting this as one possible explanation for the die-off. Another possibility is a response to toxic material that may have been dumped into the wetland around August 1989. Whatever the cause, the problem happened in the summer when the perennial plant parts were immature. This resulted in slow initial growth after breaking dormancy because the plants lacked stored energy.

The recovery of the wetland could not be attributed to the seed bank (dormant seeds buried in the marsh). If this had been the explanation, seedlings would have been observed in May 1990, and cordgrass and common three square would have been seen throughout the marsh and not at the precise locations that they had occupied before the event.

Fire

Fire contributes to the dynamics of wetlands. Forested wetlands can be subject to lightning-induced fires which might convert the forest to a wet meadow or a marsh until it slowly, over many decades, reverts to forest.

In places like the Pine Barrens of New Jersey, fire is a component of the ecosystem. Blueberries are found growing wild in low depressional areas with sparse tree cover. As the tree cover increases, blueberry bushes decrease. Periodic fires eliminate the trees in and around the blueberry bogs, opening the canopy and promoting growth of the blueberry bushes.

Periodic fires in the Florida Everglades also promote sawgrass growth and inhibit woody species

Figure 1.19 A common three square marsh island (center) surrounded by a cordgrass marsh as it appeared in 1982, one year after construction in Baltimore, Maryland.

Figure 1.20 The same marsh showing no plant life in August 1989.

Figure 1.21 The same marsh in May 1990, after upland vegetation has leafed out, but still showing no evidence of plant life.

Figure 1.22 The same marsh in August 1991 showing total recovery from the event, with the common three square marsh island in place (center left) surrounded by cordgrass marsh.

colonization, which would eventually shade out the sawgrasses.

All of the above are examples of pulsed dynamics.

Weather

When supported by an abundant supply of water wetlands are not normally severely affected by unusually wet or dry cycles in the weather. Examples of such wetlands are those supported by water from oceans, lakes, rivers, and perennial streams. Tidal freshwater wetlands, however, can be altered by extended drought periods. During these periods, freshwater flow downriver becomes limited, allowing tidal saltwater to push its way upriver, potentially killing plants that have no tolerance for salt.

Conversely, the ecologies of wetlands supported by surface water runoff or groundwater could be severely affected by unusually wet or dry cycles in the weather. Plant, animal, and invertebrate communities in these wetlands are adapted to the wetlands' normal hydroperiods with their usual fluctuations. When these hydroperiods fluctuate widely from the usual, the wetland ecologies will be changed to a degree dependent upon the duration of the fluctuations.

REFERENCES

Environmental Laboratory. 1987. *Army Corps of Engineers Wetland Delineation Manual.* U.S. Army Corps of Engineers, Waterways Experiment Station, Vicksburg, MS. Technical Report Y-87-1.

Tiner, R. W. 1998. *In Search of Swampland.* Rutgers University Press, New Brunswick, NJ.

Webster's Ninth New Collegiate Dictionary. 1989. Merriam-Webster Inc., Springfield, MA.

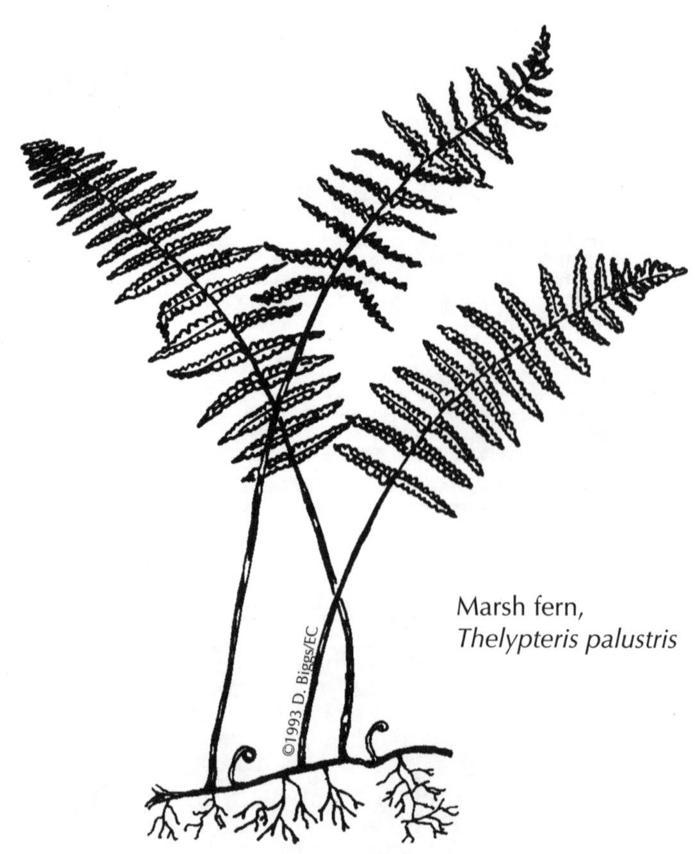

Marsh fern,
Thelypteris palustris

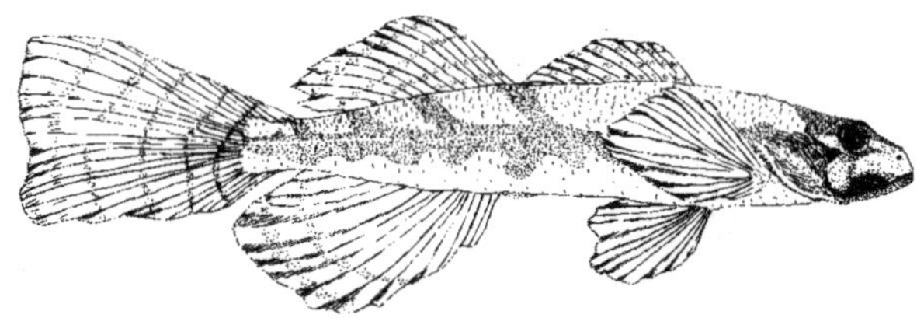

2 WATER SUPPLY

IMPORTANCE

Water is the most critical ingredient in wetlands. Without the correct quantity or supply of water for the designed wetland's hydrology (hydroperiods and water depths), the planned wetland will not be a success. The water supply is so critical that, when designing a planned wetland, it is important to be certain that the water supply is greatly in excess of what is needed. Any excess water can be discharged and returned to the water source.

There are so many different types of wetlands, each having its own unique hydrology, that the question of water supply can be a complex and challenging exercise, particularly if the supply of water is limited. Consequently, the best situation for planning a wetland is to ensure that the water supply is abundant or unlimited, remembering that any excess water can always be returned to the source.

UNLIMITED OR ABUNDANT WATER SUPPLY

Salt marshes, mangrove forests, and brackish wetlands whose water sources are the oceans can be considered to be supported by an unlimited water supply. Tidal freshwater wetlands and nontidal wetlands connected to rivers, lakes, perennial streams, and intermittent streams associated with large watersheds may be considered to be supported by an abundant water supply.

Elevations of Water Conveyance Structures

Water generally is conveyed to the planned wetland from the water body that is the source of water. If water is being conveyed to the planned wetland by means of a pipe or culvert, the pipe's or culvert's invert (bottom) often is set at the low water level in the water body; and as the water elevation increases, water is pushed through the pipe or culvert to the same increased elevation in the planned wetland.

When a planned tidal wetland is connected to tidal water by a pipe, a sufficiently large diameter pipe should be used. If the pipe is too small, the tide in the planned wetland will be restricted. A restricted tide is one that never reaches its usual high and low elevations because the water conveyance structure (channel, pipe, culvert) is too small to convey the volume of water at a rate necessary to reach these elevations. When there is a restricted tide, the high tide will be lower and the low tide will be higher than those in the tidal water body. A restricted tide should be dealt with by replacing the pipe or water conveyance structure with a substantially larger one. If this is not done, there probably will be problems identifying the proper elevations for the planned vegetation communities.

If the planned wetland is being connected directly to the water body, all of the planned wetland's other earthwork can be completed after construction of a berm separating the planned wetland and the water body. The berm keeps water out of the planned wetland during its excavation. The berm is removed during a low water elevation period in the water body (see **Figure 2.1**) so as to minimize the rate of water flow from the water body to the planned wetland and its erosional consequences.

Elevations for Vegetation Communities

Aquatic plant species can be assigned to the various planned wetland's elevations according to their tolerance for its designed hydrology (hydroperiod and depth of flooding). The hydrology information given for the plants listed in Chapter 7 may be useful in this approach to planting. However, Chapter 7 presents only a limited number of aquatic plants. Additionally, the precise elevation ranges for different vegetation communities associated with a given wetland's hydrology are

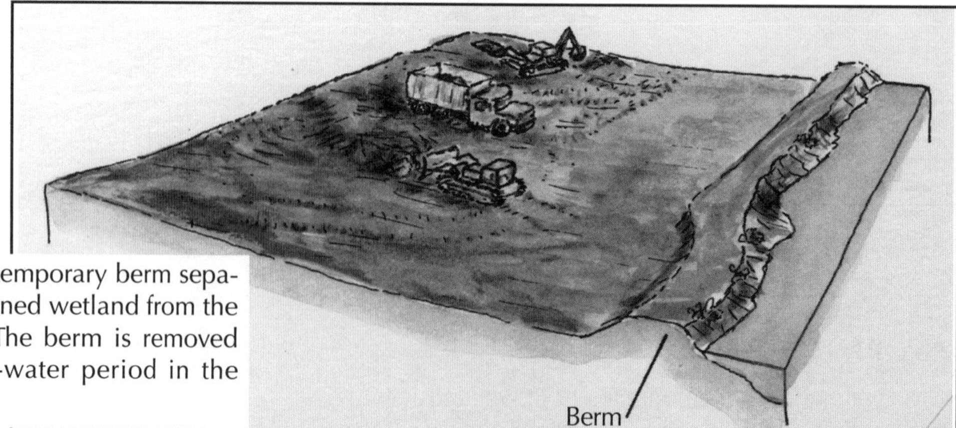

Figure 2.1 A temporary berm separating the planned wetland from the water body. The berm is removed during a low-water period in the water body.

not possible to predict. Consequently, the use of biological benchmarks can greatly expedite the planned wetland design process, as well as reliably establish the correct elevation ranges for the various vegetation communities that are to be planted.

Biological Benchmarks

A wetland biological benchmark is a plant community of one or more plant species that has for many years been thriving under the same hydrology as will be utilized at the planned wetland site. The elevation range (the high and low elevations) of the biological benchmark determines the heights at which all species of the biological benchmark will grow in the planned wetland once it has been graded and connected to the biological benchmark's hydrology. Biological benchmarks reflect:

- seasonal water levels in lakes, rivers, and streams;
- tidal information such as the normal high water and high spring tide elevations;
- maximum salt content in the water; and
- maximum depth of flooding, recorded by the silt stains on the stems and trunks of the plants.

Figure 2.2 shows surveyors taking the elevation ranges of five biological benchmarks in a nontidal wetland that will be used in the design of a neighboring planned wetland site. Biological benchmarks will not be available for:

- a planned wetland that will be excavated to intercept the water table; or
- a planned wetland that will be excavated to collect stormwater runoff from its watershed.

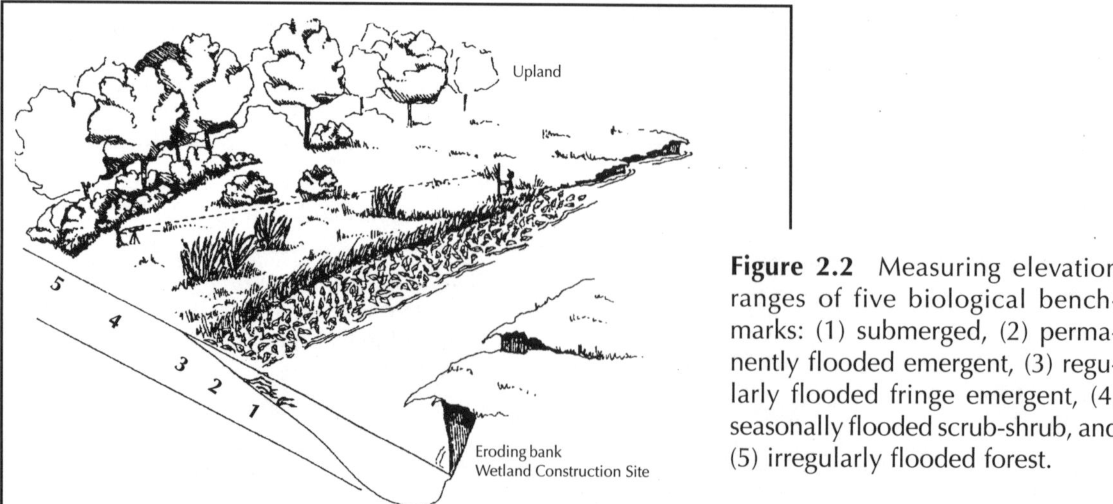

Figure 2.2 Measuring elevation ranges of five biological benchmarks: (1) submerged, (2) permanently flooded emergent, (3) regularly flooded fringe emergent, (4) seasonally flooded scrub-shrub, and (5) irregularly flooded forest.

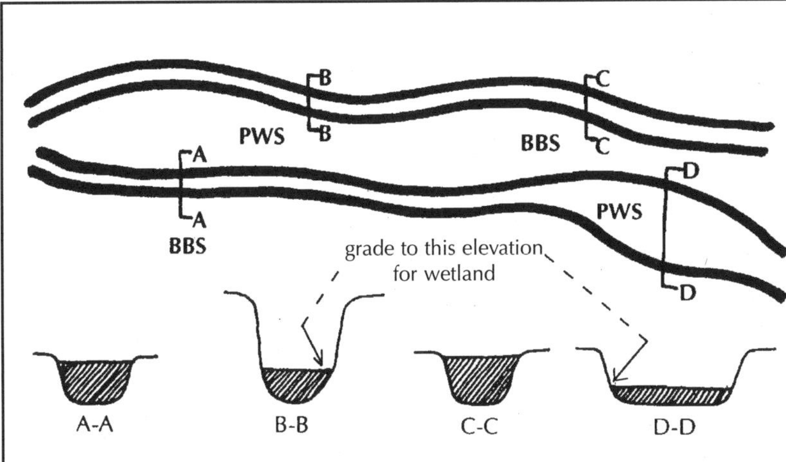

Figure 2.3 Grading a planned wetland site (PWS) that borders a stream (or river) to provide the same cross-sectional area as found at a nearby biological benchmark site (BBS) will ensure that both the PWS and the BBS will flood simultaneously as long as their associated stream (or river) sections remain stable.

This is because there are no existing wetlands that share the same hydrology.

In using biological benchmarks, it is critical to be certain that the hydrology experienced by the biological benchmark is identical to that which the planned wetland site will experience. For example:

- The planned wetland and the biological benchmark sites should be as close together as possible because their water supplies are more likely to differ if they are distant. With two greatly separated sites along a perennial stream, the downstream site will have a larger watershed and water supply than the upstream site.
- The cross-sectional area of a bank-full-stream (the area across the stream from the tops of the banks to the stream bottom) adjacent to a floodplain or riparian biological benchmark should be used to set the floodplain and bank-full grades at the planned wetland site. Doing this ensures that the bank at the planned wetland site will overflow and the floodplain will flood at the same time and in the same way as at the biological benchmark site (**Figure 2.3**).
- In tidal rivers, the tide range will increase significantly when the river narrows. For example, the tide range at the mouth of the Potomac River (7 miles wide) is 1.5 feet compared with a tide range of 2.5 feet at Washington, D.C. where the river narrows to about 3,000 feet. (See **Figure 2.4** below.) Consequently, it is important to make sure that the biological benchmark site in a tidal river is as close as possible to the planned wetland site and that the river width is not significantly different at the two locations.

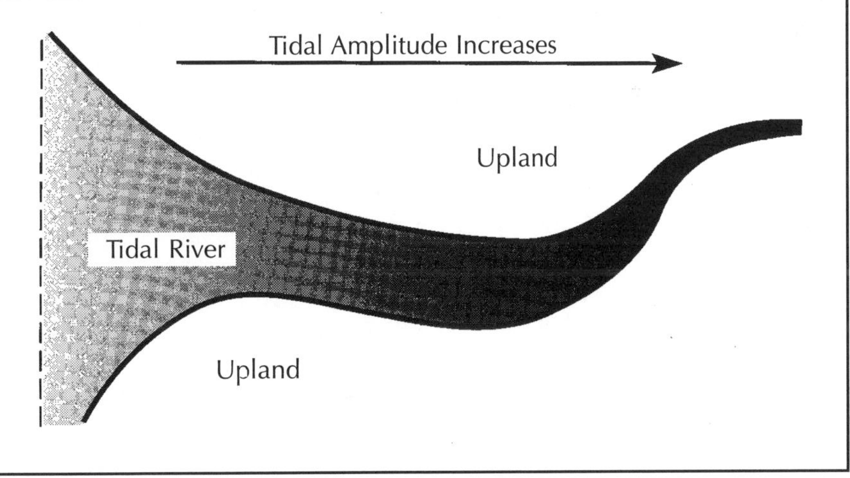

Figure 2.4 As tidal rivers narrow, the volume of water moving upriver must fit within a smaller cross section. This results in increasingly higher tides upstream.

A good check when using biological benchmarks is to obtain biological benchmark data for the same vegetation community from more than one site at differing locations. If all data are comparable, use the average value. If the data from the most distant sites are substantially different from those from the sites closest to the planned wetland, use the data from the closest sites.

Water Conveyance from Perennial Streams

In general, it is illegal to divert a stream for any reason. Before designing a planned wetland using any stream flow, check to make sure that this is a permitted activity in your region. Some states, particularly in the West, have water rights regulations which may preclude a project from even diverting flows above the base flow water level. Connections to perennial streams at their base flow elevations generally will not be permitted, as doing so may affect resources downstream. Consequently, connections to perennial and intermittent streams must be done during high flow periods. In doing so, determine first at what height above base flow the stream will provide the quantity of water needed for the planned wetland.

To determine whether a stream can provide the quantity of water for a planned wetland, four pieces of information are needed: (1) the width of the stream at the connection to the planned wetland, (2) the peak height of the stream above base flow after a given storm intensity, (3) the stream velocity at peak height, and (4) the time period during which peak flow occurs. All of this information can be collected inexpensively.

Stream Height and Stream Velocity

A rain gauge should be placed near the stream, in an open area. Place a stake marked in 0.1 foot intervals firmly in the ground at the water's edge during base flow. The zero marking should be at the base flow water surface. The length of the stake should be greater than the possible peak height of the stream. During and after a storm event, note the stream height hourly, or more frequently if the stream height is changing rapidly. Determine for a given precipitation event: (1) how quickly the peak stream height is reached, (2) how long the stream remains at different heights above base flow, including the peak height, and (3) how long it takes for the stream to return to base flow. Using the rain gauge determine the quantity of rain that has fallen in inches per hour.

Stream height and stream velocity data can be collected at the same time. Prior to the storm event, place stakes at both ends of a relatively straight stretch of the stream and measure the distance between the stakes. During the storm event, place a floating object in the water at the upstream stake and note with a stopwatch the time it takes to reach the downstream stake. Record the data as feet per second. Sticks, or even apples or oranges can be used for this purpose.

Calculating Available Water Above Base Flow

Assume that the data are as follows for a storm event that lasted 16 hours and delivered 1.2 inches of precipitation:

1. The stream width is 6.25 feet at the planned wetland connection point.
2. For three hours (or 10,800 seconds) the stream is one foot above base flow, and the velocity is 5 feet per second.
3. For two hours (7,200 seconds) the stream is two feet above base flow, and the velocity is 8 feet per second.
4. For 1.5 hours (5,400 seconds) the stream is one foot above base flow, and the velocity is 6.5 feet per second.

Therefore:

(6.25 ft)(1 ft)(5 ft/sec)(10,800 sec)
= 337,500 cubic feet

(6.25 ft)(2 ft)(8 ft/sec)(7,200 sec)
= 720,000 cubic feet

(6.25 ft)(1 ft)(6.5 ft/sec)(5,400 sec)
= 219,375 cubic feet

TOTAL WATER VOLUME ABOVE BASE FLOW
= 1,276,875 cubic feet

This is a conservative estimate since volumes were estimated for only three heights during 6.5 hours for a storm that lasted 16 hours. Furthermore, a constant stream width as a function of height was assumed. Usually, stream widths increase as the stream flow rises within its banks. However, when using this water to bring a 0.5 acre planned wetland having a 0.25 foot depth of water to its pool level (highest water level) having 0.5 foot of water, only 5,445 cubic feet of water would be needed, or only 0.43% of the total water volume estimated above.

Whatever the water conveyance mechanism (swale, pipe, etc.), it would not be possible to transfer 100% of the volume above base flow to the planned wetland at its connection to the stream. With a direct connection (no pipe or channel), possibly as much as 25% of the water above base flow may be transferred. It is always desirable to have an excess of water supplied to the planned wetland because any excess can be returned to the water body, which is in this instance the stream.

When intercepting stream flow to support a planned wetland, always complete a water budget calculation (page 38) to make certain that the water supply from the stream will exceed the wetland's needs.

Groundwater Springs

It is not possible to reliably design planned wetlands that are supported only by groundwater through the interception of the water table. This is because a water budget cannot be completed for the wetland since the water supply is so uncertain. Groundwater springs are a different situation. The groundwater discharged from springs can be seen and measured. The water supply can be quantified. A spring may have a discharge that is fairly constant, or the discharge may be seasonal or variable. Before deciding to use a spring as a source of water to support a planned wetland, make certain that the supply is more than the amount needed for the type of planned wetland that is being considered.

To measure the discharge, in gallons (or cubic feet) per day, from the spring over time, include the usual dry and wet periods. Discharged water can fill any container of known volume in a measured period of time; that data can later be converted to volume per day. Perform a water budget on the planned wetland design (page 38) and determine whether the spring supply is sufficient for the planned wetland's needs. If insufficient, determine if there are other water sources available to supplement the spring supply in order to exceed the total supply needed.

If the spring can be used, analyze the land use of the probable ground watershed to see if future development will adversely impact the available water in the aquifer and the resultant water supply from the spring. Any change that will reduce the infiltration of water to the aquifer will likely reduce the water supply from the spring. If the spring remains a likely candidate to supply water to the planned wetland, analyze the contents of the water. This is to make certain there are no substances in the water (pesticides, oil, metallic compounds, etc.) that would have a negative impact on the ecology of the planned wetland.

If these analyses show that the spring is a reliable and suitable sole or combined water source for the planned wetland, move forward with the design.

LIMITED OR UNCERTAIN WATER SUPPLY

Precipitation that falls to the earth follows a number of different pathways (**Figure 2.5**). Some of the water will flow down a drainage area and into a stream or other water body. This is called runoff. Some of the water will infiltrate the soil and increase the soil moisture. This water will then return to the atmosphere through evaporation and transpiration. Any excess soil moisture will continue to infiltrate and recharge the groundwater.

Groundwater discharge into a stream provides the base flow of the stream. Base flow varies seasonally, and sometimes dramatically, depending upon precipitation and the recharge rate of the water table. During and immediately following rain events, runoff enters streams and adds to the base flow leading to stream flow.

Wetlands supported only by runoff are those whose water supply is potentially limited because (1) the size of its watershed or drainage area may be limited, (2) the frequency and intensity of precipitation is never known before the event, and (3) the outside conditions (such as temperature, moisture content of soils, the densities and heights of plants that intercept precipitation) that affect runoff are never known before the event. Because of these factors it is impossible to predict runoff accurately and to guarantee the hydrology of a planned wetland supported only by runoff.

Wetlands supported only by groundwater through the interception of the water table are those whose water supply is potentially uncertain because (1) unlike the surface watershed, the ground watershed is difficult to locate, (2) the volume of water in the ground watershed is not known, (3) the rate of discharge of groundwater into an excavation is not known, and (4) the rate of recharge of the groundwater is not known. Because of the above,

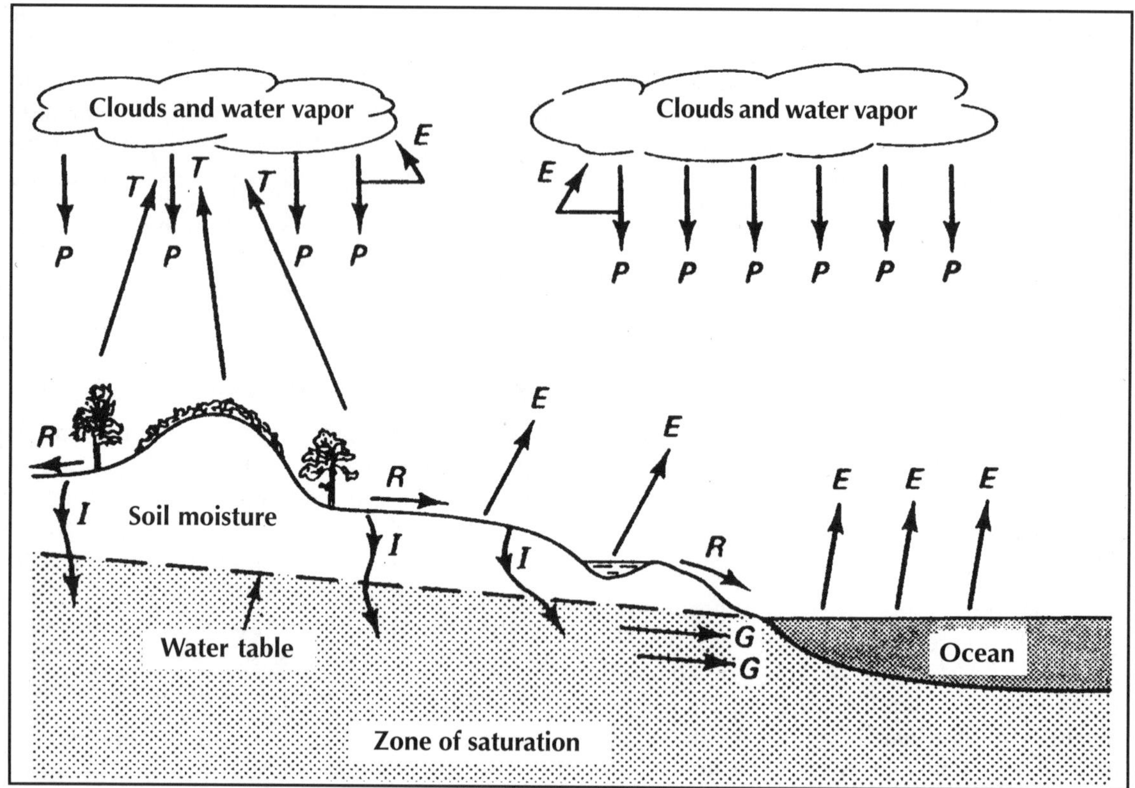

Figure 2.5 Regional water cycle where E = evaporation, G = groundwater flow, I = infiltration, P = precipitation, R = runoff, and T = transpiration.

it is impossible to predict accurately the groundwater discharge into a wetland and to guarantee the hydrology of a planned wetland supported only by groundwater discharge.

Hydrology is not an exact science. Reasonable well-founded assumptions are necessary to solve practical problems. This is especially true when calculating both runoff and groundwater discharge into excavations.

Defining a Drainage Area

Unless the intention is to intercept stream flow, as discussed earlier, or to otherwise collect and convey runoff to a planned wetland site, it is recommended that the wetland be placed within its drainage area or watershed. Topographic contour lines indicate the watershed's boundary. Remember that contour lines that point uphill define a swale and contour lines that point downhill define a ridge.

The place where excess water is released from the wetland is the **discharge point**. Start at the discharge point to define the planned wetland's drainage area. The contours nearest the discharge point indicate two opposite boundaries of the planned wetland's drainage area. Draw the boundary on one side, starting at the discharge point. Move uphill from contour line to contour line by the shortest possible path. Cross each contour line at a right angle. The line that you are drawing is a ridge line. On the planned wetland's side of the line, water will flow to the discharge point. On the other side of the line, water will flow away from the wetland.

As the topographic high point is approached, stop working on that line. Start again at the discharge point, going uphill on the other side of the discharge point. If a connection is made at the high point with the first line, the drainage area boundary for the planned wetland is complete. If the lines do not connect, then the lines have reached two separate topographic high knobs, which then must be connected through what is called a **saddle point** (Ferguson and Debo 1990).

Figure 2.6 shows a planned wetland that has been placed in part of a watershed, with its drainage area defined as described above.

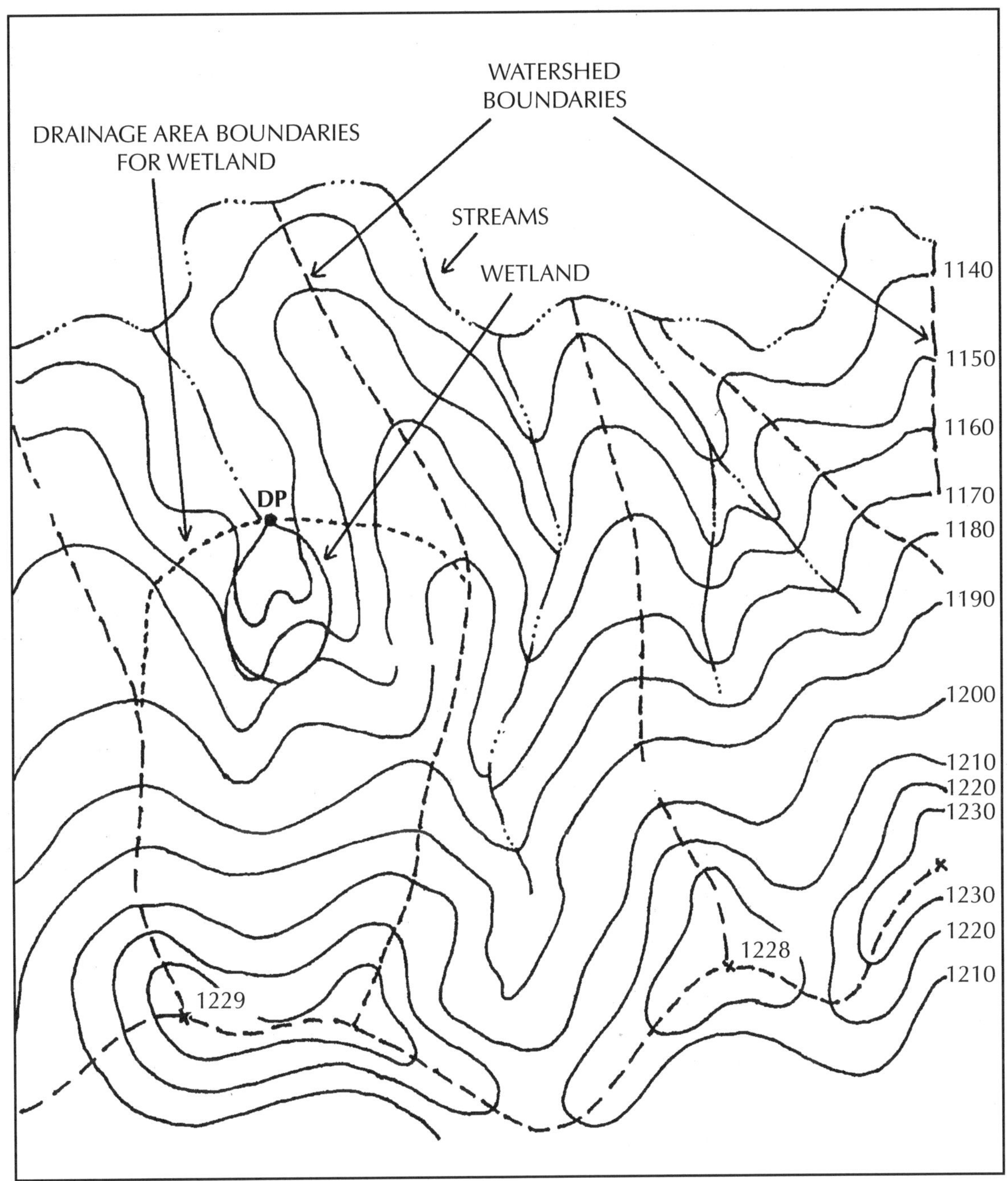

Figure 2.6 Defining the drainage area for a planned wetland that is placed within its own drainage area. Scale: 1 inch = 2,000 feet.

The Water Budget

The water budget for a planned wetland considers all water inputs to and outputs from a wetland. It includes surface water and groundwater components. It is given by **Equation 2.1**.

$$\Delta S = P + R_{in} + W_g - W_{out} - I - ET \qquad \textbf{(2.1)}$$

Where:

ΔS = change in water storage in the wetland.

P = precipitation falling directly on the planned wetland. When the planned wetland is placed in its own drainage area, as in **Figure 2.6**, P is taken out of **Equation 2.1** and it is used only to calculate R_{in}.

R_{in} = runoff to the wetland, or the discharge point, as in **Figure 2.6**. This includes I, ET, and P components.

W_g = groundwater discharge into the wetland.

W_{out} = water leaving the wetland, or the discharge point as in **Figure 2.6.**

I = infiltration to soils and groundwater.

ET = evapotranspiration or the consumptive use.

When a wetland is supplied with water by intercepting stream flow above base flow, R_{in} refers to what is intercepted. Since base flow is not being intercepted, W_g in **Equation 2.1** is zero.

In **Equation 2.1**, it is assumed that ET ≅ E (evaporation) (Fetter 1998). Consequently, E is not included in the water budget, and the water loss due to evaporation and/or transpiration year-round is estimated by the calculation of ET. If the wetland contains only open water, the estimated ET can be assumed to equal E. Fetter (1998) states that the water loss from a pond is about the same, whether or not emergent aquatic vegetation is present. Additionally, measurement of monthly turf evapotranspiration and lake evaporation for cities throughout the United States that are provided in the Water Balance Data tables in Ferguson and Debo (1990) are frequently approximately the same.

The consumptive use coefficient (Viessman et al. 1989) for rice is used as an approximation of ET for wetlands. Others (Penman 1948 and Criddle 1958) have equated ET and consumptive use. **Equation 2.2** is used to calculate ET.

$$ET = 1.2\ p\ t \div 36{,}000 \qquad \textbf{(2.2)}$$

Where:

ET = the daily loss of water through evaporation or evapotranspiration measured in feet of water depth.

1.2 = the consumptive use coefficient for rice in inches of water depth per month.

p = the average monthly daytime, for the months during the growing season, given as a percentage of the year, obtained from **Table 1**.

t = the average monthly temperature in °F from local or other sources for the growing season.

36,000 = the conversion factor to change from inches per month to feet per day, times 100. The units are days per month, inches per foot.

Infiltration is that process by which precipitation moves downward by gravity through the ground, replenishing soil moisture and recharging groundwater. It is solely responsible for the base flows in streams during dry periods. Infiltration is a complex process and cannot yet be estimated through calculations. The best way to estimate it is through empirical observations. It is influenced by the type and extent of vegetative cover, soil type, soil moisture, soil temperature, the rate of water supply, and water quality.

Infiltration rate at the substrate surface is highly dependent on the condition of the surface. Soil compaction can largely limit infiltration. The inwash of fine-grained soils may seal the surface so the infiltration rates become low or vanish, even when the underlying soils are highly permeable.

The expected infiltration rates for all of the Soil Conservation Service's (SCS) uncompacted soil texture classes are listed in **Table 2**. These can be used in calculating the water budget for a planned wetland. Field estimates of infiltration rates at a planned wetland site can be obtained rapidly by using a double ring infiltrometer, which is available commercially or might be borrowed from an engineer or from a geotechnical firm.

Wetlands Supported by Stormwater Runoff

If the plan is to construct a wetland that is supported only by stormwater runoff, the water budget (**Equation 2.1**) becomes **Equation 2.3**.

$$\Delta S = R_{in} - W_{out} - I - ET \qquad \textbf{(2.3)}$$

As the wetland is part of its drainage area (**Figure 2.6**), P (precipitation) is removed from the water

Table 1 Monthly Average Daytime as a Percentage of Yearly Daytime

Latitude	Jan.	Feb.	March	April	May	June	July	Aug.	Sept.	Oct.	Nov.	Dec.
NORTH LATITUDE												
60	4.67	5.65	8.08	9.65	11.74	12.39	12.31	10.70	8.57	5.98	5.04	4.22
50	5.98	6.30	8.24	9.24	10.68	10.91	10.99	10.00	8.46	7.45	6.10	5.65
40	6.76	6.72	8.33	8.95	10.02	10.08	10.22	9.54	8.39	7.75	6.72	6.52
35	7.05	6.88	8.35	8.83	9.76	9.77	9.93	9.37	8.36	7.87	6.97	6.86
30	7.30	7.03	8.38	8.72	9.53	9.49	9.67	9.22	8.33	7.99	7.17	7.15
25	7.53	7.14	8.39	8.61	9.33	9.23	9.45	9.09	8.32	8.09	7.40	7.42
20	7.74	7.25	8.41	8.52	9.15	9.00	9.25	8.96	8.30	8.18	7.58	7.66
15	7.94	7.36	8.43	8.44	8.98	8.80	9.05	8.83	8.28	8.26	7.75	7.88
10	8.13	7.47	8.45	8.37	8.81	8.60	8.86	8.71	8.25	8.34	7.91	8.10
0	8.50	7.66	8.49	8.21	8.50	8.33	8.50	8.49	8.21	8.50	8.22	8.50
SOUTH LATITUDE												
10	8.66	7.87	8.53	8.09	8.18	7.86	8.14	8.27	8.17	8.62	8.53	8.88
20	9.24	8.09	8.57	7.94	7.85	7.43	7.76	8.03	8.13	8.76	8.87	9.33
30	9.70	8.33	8.62	7.73	7.45	6.96	7.31	7.76	8.08	8.97	9.24	9.85
40	10.27	8.63	8.67	7.49	6.97	6.37	6.76	7.41	8.02	9.21	9.71	10.49

budget formula. Calculations will be for runoff (R_{in}) to the discharge point in **Figure 2.6**. Groundwater discharge (W_g) to the wetland also is removed from the water budget, since the wetland is supported only by runoff.

Before calculating runoff, it is important to determine whether or not all of the runoff in the drainage area runs to the discharge point. If some of the runoff is conveyed by pipe or ditch to a place outside of the drainage area, then determine the square footage of the area whose runoff does not go to the discharge point. This runoff must be excluded from the calculation for runoff going to the discharge point. Water from parking lots and athletic fields will have to be looked at carefully. These areas often have a ridge in the middle so that water drains to their sides. Also look at roof drains or gutters on the buildings and adjust the square footage if some fraction of these does not drain to the discharge point.

Before calculating runoff, also examine areas outside the drainage area and determine if runoff from any of these areas is conveyed by pipe or ditch (swale) to the drainage area. If so, the square footage of these areas must be determined and their land use noted so that this runoff can be added to the calculated runoff going to the discharge point.

In order to determine whether the wetland being planned is likely to be supported by runoff from the site, the following information is needed: (1) the planned wetland's drainage area (**Figure 2.6**), (2) the soil types throughout the drainage area, including those at the planned wetland site, (3) the various land uses throughout the drainage area, (4) the average monthly rainfall during the growing season, and (5) the average monthly temperatures during the growing season.

The next step is to calculate R_{in} during the rain event. Assume that the rain event is brief enough

Table 2 Infiltration Rates for Soil Conservation Service Soil Texture Classes (from Rawls et al. 1982)

SCS Soil Texture Class	Infiltration Rate (ft/day)
sand	16.54
loamy sand	4.82
sandy loam	2.04
loam	1.04
silt loam	0.54
sandy clay loam	0.34
clay loam	0.18
silty clay loam	0.12
sandy clay	0.10
silty clay	0.08
clay	0.04

that I and ET in **Equation 2.3** will approximate zero. Therefore **Equation 2.3** simplifies to **Equation 2.4**.

$$\Delta S = R_{in} - W_{out} \quad \textbf{(2.4)}$$

Where:

ΔS = the volume of water needed to fill the planned wetland to its pool level.

Whether designing a permanently flooded or irregularly flooded planned wetland, always include a discharge structure in the design. Whether a pipe or a weir, the structure's bottom elevation will be the pool elevation. Pool elevation must be designed carefully. For an irregularly flooded wetland, the pool depth may be only one to two inches. If, on trial calculations of R_{in}, the result is always $\Delta S > 0.5\ R_{in}$, redesign the planned wetland by reducing its size and/or reducing its depth(s) of water so that $\Delta S \leq 0.5\ R_{in}$.

If the planned wetland is at its pool level (highest water elevation) when the rain starts, then ΔS is zero. If the planned wetland does not fill to its pool level during the rain event, then $\Delta S = R_{in}$. If the rain event fills the planned wetland to its pool level and excess water runs out, i.e., $W_{out} > 0$, then $\Delta S < R_{in}$. The desired result is:

$$\Delta S \leq 0.5\ R_{in}$$

This means that there will be a substantially greater water supply than needed.

At this point the type of planned wetland that will be constructed, restored, or enhanced has been decided. Its designed hydrology (hydroperiod and water depth) and its size have been determined. R_{in} can now be calculated using **Equation 2.5**.

$$R_{in} = a\ c\ i\ (3{,}630 \text{ cubic feet/acre inch}) \quad \textbf{(2.5)}$$

Where:

R_{in} = runoff in the drainage area containing the planned wetland which discharges excess water at the discharge point (**Figure 2.6**).

a = the drainage area in acres including the wetland and the discharge point.

c = the applicable runoff coefficient.

i = the average two-week precipitation during the growing season, assuming that it all comes within a 24-hour period.

Equation 2.5 is used in the Rational Method for estimating runoff from large storm events (e.g., the 50-year or 100-year storm). Consequently, the runoff coefficients are higher than those needed for the small (two-week) storm event that is being used. The runoff coefficients in **Table 4** (or the Applicable Runoff Coefficients) are reduced by 35% if $i \leq 2$ inches and reduced by 20% if $i > 2$ inches but ≤ 3 inches.

It is very important to realize that the Rational Method is imprecise for large storm events, and is even more so for the small storm events considered here. For the sake of simplicity the assumption is made that the average two-week precipitation (the average monthly precipitation during the growing season, divided by two) falls within a 24-hour period. It must be remembered that this is only an assumption. Should the average two-week precipitation fall over a three-day period, there may be no runoff. Because of the assumptions and the qualitative nature of the Rational Method, always undersize the planned wetland so that there will always be a greater water supply than needed.

To estimate the Applicable Runoff Coefficient, add together the products of each coefficient for each land use and its percentage of the total drainage area, remembering that the planned wetland is part of the drainage area and itself has a runoff coefficient. For example, if the drainage area consists of 25% land use X, 70% land use Y, and 5% wetland, the Applicable Runoff Coefficient, c, is given by:

$$c = (0.25)(c_x) + (0.7)(c_y) + (0.05)(c_{wetland})$$

In calculating R_{in}, first calculate the square footage of each land use/cover found in the drainage area, based on **Table 4**. **Table 3** lists the Hydrologic Soil Groups needed for **Table 4**. In order to assign the proper Hydrologic Soil Group for each land use/cover, soil borings will have to be taken at each land use/cover that isn't surfaced. To determine the Soil Group for the planned wetland, a soil sample will have to be obtained at the planned wetland's final grade level, not at the ground surface.

Sizing a Planned Wetland

The following steps are recommended when sizing a planned wetland:

- Select a desired size in square feet for the planned wetland.
- State the planned wetland's designed hydrology (hydroperiods and water depths).
- Place the planned wetland in the available watershed with its overflow (pool level control) at the Discharge Point as in **Figure 2.6**.
- Calculate R_{in} from **Equation 2.5**.
- Calculate ΔS from **Equation 2.4**. If $\Delta S \geq 0.5 R_{in}$, reduce the size and/or the water depth(s) of the planned wetland until $\Delta S \leq 0.5 R_{in}$. When this condition is achieved, there can be some confidence that the water supply is greater than needed to maintain the planned wetland's pool elevation after each storm event.

For an example refer to the *Sizing Exercise* below.

SIZING EXERCISE

A school wants to construct a permanently flooded wetland consisting of 50% open water having a maximum water depth of three feet, and 50% marsh that has a maximum water depth of one foot, in a schoolyard that has a three-acre watershed (drainage area). The soils in the watershed and at the final grades of the wetland are silt loam (Hydrologic Soil Group B). A wetland having a size of 0.2 acre will be tested. The watershed consists of 30% impervious roofs and pavements, 63% mowed fields with >75% grass cover, and

Table 3 Characteristics of the Hydrologic Soil Groups

Group	Characteristics
A	Sand, loamy sand, or sandy loam. These soils have low runoff potential and high infiltration rates even when thoroughly wetted. They consist mainly of deep, well drained to excessively drained sands or gravels. Infiltration rates exceed 2.04 feet per day.
B	Silt loam or loam soil. These soils have moderate infiltration rates when thoroughly wetted. They consist mainly of moderately deep to deep, well drained soils with moderately fine to moderately course texture. Their infiltration rates range from 0.54 to 1.04 feet per day.
C	Sandy clay loam. These soils have low infiltration rates when thoroughly wetted, and consist mainly of soils with a layer that impedes downward movement of water. Soils are moderately fine to fine textured. Infiltration rate is 0.34 feet per day.
D	Clay loam, silty clay loam, sandy clay, silty clay, clay (listed in order of decreasing infiltration). Soils have high runoff potential. They have very low infiltration rates when thoroughly wetted. These are mainly clay soils with high swelling potential, soils with a permanently high water table, soils with a clay pan or clay layer at or near the surface, and shallow soils over nearly impervious materials. Infiltration ranges from 0.04 foot per day for clay to 0.18 foot per day for clay loam.

Table 4 Runoff Coefficients (c) for the Rational Method

Land Use/Surface Cover	Hydrologic Soil Group			
	A	B	C	D
FOREST:				
no grazing; litter and brush cover soil	.05	.10	.30	.40
grazing; some litter covers soil	.07	.16	.35	.43
heavy grazing; no litter and brush cover soil	.08	.25	.40	.49
BRUSH, GRASS, AND WEEDS IN HUMID REGIONS:				
>75% ground cover (litter, grass and brush)	.05	.09	.23	.35
50% to 75% ground cover	.06	.11	.30	.40
<50% ground cover	.08	.26	.40	.49
DESERT SHRUB (SALTBUSH, MESQUITE, ETC.):				
>70% ground cover (liter, grass, and brush)	.09	.28	.43	.50
30% to 70% ground cover	.10	.33	.48	.55
<30% ground cover	.21	.40	.52	.62
MEADOW:				
continuous grass, not grazed, generally mowed for hay	.06	.14	.33	.40
LAWNS OR TURF:				
>75% grass cover	.07	.18	.36	.44
50% to 75% grass cover	.09	.29	.43	.50
<50% grass cover	.28	.43	.55	.65
PASTURE, GRASSLAND/RANGE IN HUMID REGIONS:				
>75% ground cover and light graze	.07	.18	.36	.44
50% to 75% grass cover	.09	.29	.43	.50
<50% grass cover	.28	.43	.55	.65
CULTIVATED LAND:				
with conservation treatment	.19	.32	.40	.46
without conservation treatment	.33	.46	.62	.71
FARMSTEADS:				
buildings, lanes, driveways, and lots	.15	.36	.47	.55
FALLOW:				
≥20% residue cover	.36	.49	.62	.68
<20% residue cover	.39	.52	.68	.78
bare soil	.40	.55	.71	.81
NEWLY GRADED AREAS:				
pervious with no vegetation	.40	.55	.71	.81
STREETS, ROADS, BUILDINGS, STRUCTURES:				
dirt roads, including right-of-way	.34	.47	.58	.65
gravel roads, including right-of-way	.39	.52	.65	.71
paved roads with open ditches, including right-of-way	.49	.65	.74	.78
impervious roofs and pavements	.94	.94	.94	.94
WATER BODIES, INCLUDING PLANNED WETLAND:				
unvegetated	1.00	1.00	1.00	1.00
vegetated	.84	.84	.84	.84

7% wetland. The mean monthly temperature and precipitation during the growing season at the site are 75 °F and 4.4 inches, respectively. The applicable runoff coefficient is:

$$c = (0.30)(0.94) + (0.63)(0.18) + (0.035)(1.00) + (0.035)(0.84) = 0.46$$

Initially, all of the wetland is unvegetated. However, since 50% of the wetland will be vegetated eventually, this is what was used above to calculate the applicable runoff coefficient. Additionally, because the mean two-week precipitation is 2.2 inches during the growing season, the applicable runoff coefficient is reduced by 20%:

$$c = 0.46 - (0.20)(0.46) = 0.37$$

From **Equation 2.5**,

$$R_{in} = (3 \text{ acres})(0.37)(2.2 \text{ inches})(3{,}630 \text{ ft}^3/\text{acre inch}) = 8{,}864 \text{ ft}^3$$

The total volume of the wetland at its pool level is 0.1 acre at 3 feet depth (4,356 ft^2 X 3 ft = 13,068 ft^3) and 0.1 acre at 1 foot depth (4,356 ft^2 X 1 ft = 4,356 ft^3) or a total volume of 17,424 ft^3. Consequently the precipitation event supplied only half the volume of water necessary to fill the wetland to its pool level. This leads to $\Delta S = R_{in}$ in **Equation 2.4**, since $W_{out} = 0$. We are looking for an abundance of water so that $\Delta S \leq 0.5\ R_{in}$. In order to meet this condition, W_{out} must be $\geq 0.5\ R_{in}$. In order for $R_{in} = W_{out}$, the wetland's area has to be reduced by at least one half. In order for $W_{out} \geq R_{in}$, the wetland's area again must be reduced by at least one half, making the new size of the wetland no greater than (0.25)(0.2 acre) or 0.05 acre (2,178 ft^2).

The size of the wetland is reduced to 2,000 ft^2. Doing this leads to an unadjusted runoff coefficient of 0.42. Reducing this by 20% leads to an adjusted runoff coefficient of 0.37.

Pine Barrens treefrog, *Hyla andersoni*

$$R_{in} = (3 \text{ acres})(0.37)(2.2 \text{ inches})(3{,}630 \text{ ft}^3/\text{acre in}) = 8{,}864 \text{ ft}^3$$

The volume of the size-adjusted wetland is 4,000 ft^3 (1,000 ft^3 + 3,000 ft^3). Consequently, W_{out} = 4,146 ft^3, leading to ΔS in **Equation 2.4** = 4,000 ft^3, which is $\leq 0.5\ R_{in}$. If this planned wetland is partially filled when it rains, and the rain event is as assumed, then ΔS will always be less than 4,000 ft^3.

Ensuring the Designed Hydroperiod

What was designed above is a permanently flooded wetland having vegetation at a maximum water depth of one foot. There is reasonable confidence that average two-week precipitation events will bring the wetland to its pool level. However, it is not known if the wetland will remain flooded between the two-week rain events. In order to determine this, a water budget must be completed for the planned wetland.

The water budget, **Equation 2.1**, simplifies to **Equation 2.3** for planned wetlands placed within a watershed (drainage area) with runoff as the only water source. Following a normal two-week precipitation event that fills the wetland to its pool level, determine the water loss until the next two-week precipitation event, thus (**Equation 2.3** simplifies to **Equation 2.6**):

$$\Delta S = -I - ET \qquad \textbf{(2.6)}$$

Infiltration rates, I, are given in **Table 2** and evapotranspiration rates are given using Equation 2.2. In the *Sizing Exercise*, the soils throughout the drainage area and at the final grades of the planned wetland are silt loam. From **Table 2**, water is lost from the planned wetland due to infiltration at a rate of 0.54 foot per day. Using a latitude of 40° in **Table 1** and 75° F as the average summer temperature from April through September, ET is calculated as 0.02 foot/day. During the two-week (14-day) period following the precipitation event, ΔS in **Equation 2.6** equals -9.46 feet, 9.18 coming from infiltration and 0.28 coming from evapotranspiration.

Clearly the planned wetland is losing water through infiltration at too fast a rate to be a permanently flooded wetland. If the infiltration can be reduced to nearly zero, the designed hydrology can be realized.

Methods of eliminating or reducing infiltration include:

1. Line the site with clay, then topsoil, in areas where vegetation is desired.
2. Compact the soil with a roller or vibrating compactor until the desired infiltration is achieved, as determined by using an infiltrometer. Areas where vegetation is desired should be covered with topsoil following compaction.
3. Line the site with polyvinyl chloride (PVC) having at least a 20 mil thickness. The entire liner should be covered with at least one foot of soil. Contact the Watersaver Company, Inc., P.O. Box 16465, Denver, CO 80216 for more information.
4. Line the site with a geosynthetic clay liner, such as Claymax 200R. Cover the liner with at least one foot of soil. Contact James Clem Corp., 444 North Michigan, Suite 1610, Chicago, IL 60611 for more information.

Items 1, 3, and 4 above are the best methods for eliminating infiltration.

If you are willing to wait for up to a year, natural sealing of wetlands having a silt loam soil can occur. Natural nutrient cycling in wetlands due to plant biomass decomposition often leads to algal blooms during the growing season. As the algae die and settle to the bottom of the wetland, the pore spaces in the soil are sealed, thereby eliminating infiltration. It has been reported that adding fertilizer to the water in the wetland will accelerate the sealing process. Inflow of silts to the wetland during rain events also may lead to sealing of the pore spaces in the soil.

Wetlands Supported by Groundwater

As discussed earlier, it is difficult, if not impossible to design with precision wetlands that are supported solely by groundwater. When (1) confident that the groundwater supply is abundant, (2) designing a small[1] wetland (<2000 ft^2), or (3) designing a wetland where groundwater will be augmented by another water source, the water table will need to be monitored in the area were the planned wetland will be placed. To monitor the water table, several monitoring wells or piezometers will have to be installed to determine the groundwater levels for a small wetland.

A piezometer is a shallow well, usually made of PVC pipe, for monitoring groundwater levels. In order to determine where the high water table will be encountered, dig a hole during the dry season when the water table is expected to be at its lowest. Use a post hole digger or a soil auger and excavate one or two holes until groundwater is encountered. Obtain from a local hydrogeology firm piezometers long enough to extend from three feet above the ground level to several inches below the low water table. Insert each piezometer into a hole until the bottom of the screened portion of the piezometer is just below the low water table. Backfill each hole as shown in **Figure 2.7**.

To monitor the well, unscrew the cap and put an appropriate measuring device (e.g., a meterstick) into the pipe. Record the distance between the top of the pipe and the groundwater surface. To get the depth to groundwater, subtract the distance from the top of the pipe to the ground surface from the reading. Daily monitoring will give the best information; but, at a minimum, readings should be obtained weekly.

Monitoring should continue so as to include the times of high and low water tables for the year. A rain gauge at the site will be helpful in order to record the water table's response to rain events. By graphing the water table as a function of time, the relationship between when and where the water table is on the site and its response to rain events will be evident.

In a seasonally inundated wetland, the final grades of the wetland would be just below the high water table elevation so that the planned wetland's soils would be saturated during the early to late spring (in the East) and dry during the rest of the growing season. In a permanently inundated wetland, the final grades of the planned wetland would be at or just below the elevation of the low water table. In this design, a water control structure should be part of the plan so that the water depth can be controlled and any excess water returned to a waterway.

---

[1] Large excavations of several acres or more into the water table are known to influence the water table profiles (Winston 1996, 1997; Garbisch 1994). Consequently, to minimize this impact, excavations should be kept small (< 2000 ft^2).

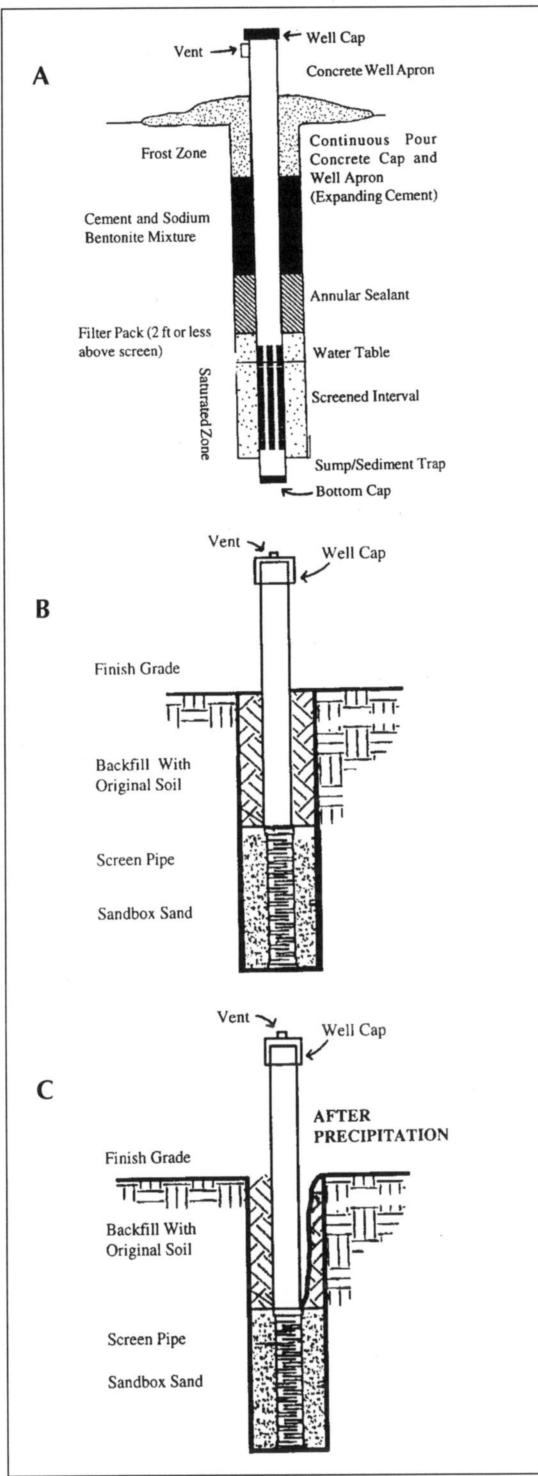

Figure 2.7 Backfill the piezometer as shown in **A.** Do not backfill with original soil as shown in **B**, as erosion is likely upon contact with surface water. This will lead to surface water running down the side of the well, as shown in **C**, and would invalidate the monitoring results.

In many regions of the country a permit must be obtained for each monitoring well that is installed, and monitoring reports must be submitted to the permitting agency. The permits should be easy to obtain, because a planned wetland will be very small and not affect the groundwater supply for other permitted purposes.

REFERENCES

Criddle, W.D. 1958. "Methods of Computing Consumptive Use of Water" in *Proceedings of the American Society of Civil Engineers Journal, Irrigation Drainage Division;* 84 (IRI): 1-27.

Ferguson, B. and T.N. Debo. 1990. *On-Site Stormwater Management (Second Edition).* Van Nostrand Reinhold, New York, NY. 270 pp.

Fetter, C.W. 1998. *Applied Hydrology (Second Edition).* Merrill Publishing Company, Columbus, OH. 592 pp.

Garbisch, E.W. 1994. *Achieving the Correct Hydrology to Support Constructed Wetlands.* Environmental Concern Inc., St. Michaels, MD. 90 pp.

Penman, H.L. 1948. "Natural Evaporation from Open Water, Bare Soil, and Grass" in *Proceedings of the Royal Society of London, Series A.* 193(1032): 120-145.

Rawls, W.J., D.L. Brakensick, and K.E. Saxton. 1982. *Transactions of the American Society of Agricultural Engineers.* 25(5): 1316-1320, 1328

Viessman, W. Jr., G.L. and J. W. Knapp. 1989. *Introduction to Hydrology (Third Edition).* Harper Collins Publishers, New York, N.Y. 780 pp.

Winston, R.B. 1996. "Design of an Urban Ground-Water-Dominated Wetland." *Wetlands* 16(4): 524-531.

Winston, R.B. 1997. "Problems Associated with Reliably Designing Groundwater-Dominated Wetlands." *Wetland Journal* 9(1): 21-25.

3 PERMITS

Before designing a planned wetland, check to see if any permits will be required. In order to protect wetland resources, the Federal government and many state and local governments require people to obtain permission to perform projects in wetlands. Permission is granted by issuing a permit or a letter of authorization specifying what activities are allowed in the wetland. The Federal government regulates wetlands under Section 404 of the Clean Water Act. The U.S. Army Corps of Engineers (COE) is the Federal agency authorized to issue permits for activities affecting wetlands. Often the COE works jointly with state agencies to issue permits. Local, county, and city governments may also require permission through a permitting process or special authorization.

A wetland restoration or enhancement project will most likely require a permit or other type of authorization. Creating a wetland does not generally require a permit unless the wetland will be created in an area already exhibiting wetland characteristics (such as soils and vegetation). In most cases, government authorities will help submit all the necessary information and adjust the project, if necessary, to be in compliance with the law. For smaller projects, regulatory agencies may only need to be informed of the activity. Time and money will be saved by designing a project that do not require a full permit application. Generally speaking, projects affecting less than one-half of an acre of wetlands will not require a full permit application, although projects larger than one-tenth of an acre may require review. Activities that typically require permits include clearing vegetation in a wetland in order to plant, digging or ditching in wetlands (to grade or to create a ponded area, for example), and placing structures in wetland areas (boardwalks, overlooks).

TYPES OF PERMISSION

Clean Water Act Section 404 Permits are issued by the COE with oversight by the Environmental Protection Agency for major projects destroying or affecting wetlands. Activities that are regulated include dredging or filling of wetlands for roads, airports, dams, and commercial or residential development. Both individual and nationwide permits are issued. Individual permits are the most expensive and time consuming to obtain, but are not likely to be needed for a schoolyard wetland.

Nationwide Permit

Nationwide permits are issued for specific types of activities having relatively minor impacts on wetlands. These activities do not require further COE approval as long as the conditions set in the nationwide permit are met. As of June 2000 there were forty-four nationwide permit categories (*Federal Register,* Vol. 65, No. 47, March 9, 2000). Many nationwide permits only require that the applicant notify the COE about the project.

Under the nationwide permit program, the COE may require written notification. The notification should contain the name, address, and telephone numbers of the applicant, the location of the proposed project, and a brief description of the project (purpose, list of any direct or indirect adverse environmental impacts, and other nationwide permits being applied for in connection with the project).

General Permit

Some states have their own versions of the nationwide permit program approved by the COE. These individual states may specify certain types of wetland activities that may occur with minimal authorization, usually a general wetland permit. To obtain a general permit, project leaders must submit an application to the state.

Letter of Authorization

For a project or activity that will not have a significant wetland impact, but still requires regulatory approval, the COE and many states provide a letter of authorization. This requires less time and effort to obtain than a general permit.

NATIONWIDE PERMIT #27

Stream and Wetland Restoration Activities is the most common type of nationwide permit sought for restoration or enhancement projects. This permit will cover most, if not all, schoolyard wetlands. The permit states:

> 27. *Stream and Wetland Restoration Activities*. Activities in waters of the United States associated with the restoration of former waters, the enhancement of degraded tidal and nontidal wetlands and riparian areas, the creation of tidal and nontidal wetlands and riparian areas, and the restoration and enhancement of nontidal streams and nontidal open water areas as follows:
>
> (a) The activity is conducted on:
>
> (1) Non-Federal public lands and private lands, in accordance with the terms and conditions of a binding wetland enhancement, restoration, or creation agreement between the landowner and the U.S. Fish and Wildlife Service (FWS) or the Natural Resources Conservation Service (NRCS) or voluntary wetland restoration, enhancement, and creation actions documented by the NRCS pursuant to NRCS regulations; or
>
> (2) Any Federal land; or
>
> (3) Reclaimed surface coal mined lands, in accordance with a Surface Mining Control and Reclamation Act permit issued by the Office of Surface Mining or the applicable state agency (the future reversion does not apply to streams or wetlands created, restored, or enhanced as mitigation for the mining impacts, nor naturally due to hydrologic or topographic features, nor for a mitigation bank); or
>
> (4) Any private or public land;
>
> (b) *Notification:* For activities on any private or public land that are not described by paragraphs (a)(1), (a)(2), or (a)(3) above, the permittee must notify the District Engineer [COE] in accordance with General Condition 13; and
>
> (c) Only native plant species should be planted at the site, if permittee is vegetating the project site.
>
> Activities authorized by this NWP [Nationwide Wetland Permit] include, but are not limited to: the removal of accumulated sediments; the installation, removal, and maintenance of small water control structures, dikes, and berms; the installation of current deflectors; the enhancement, restoration, or creation of riffle and pool stream structure; the placement of in-stream habitat structures; modifications of the stream bed and/or banks to restore or create stream meanders; the backfilling of artificial channels and drainage ditches; the removal of existing drainage structures; the construction of small nesting islands; the construction of open water areas; activities needed to reestablish vegetation, including plowing or discing for seed bed preparation; mechanized land clearing to remove undesirable vegetation, and other related activities.
>
> This NWP does not authorize the conversion of a stream to another aquatic use, such as the creation of an impoundment for waterfowl habitat. This NWP does not authorize stream channelization. This NWP does not authorize the conversion of natural wetlands to another aquatic use, such as creation of waterfowl impoundments where a forested wetland previously existed. However, this NWP authorizes the relocation of nontidal waters, including nontidal wetlands, on the project site provided there are net gains in aquatic resource functions and values. For example, this NWP may authorize the creation of an open water impoundment in a nontidal emergent wetland, provided the nontidal emergent wetland is replaced by creating that wetland type on the project site. This NWP does not authorize the relocation of tidal waters or the conversion of tidal waters, including tidal wetlands, to other aquatic uses, such as the conversion of tidal wetlands into open water impoundments.
>
> *Reversion*. For enhancement, restoration, and creation projects conducted under paragraphs (a)(2) and (a)(4), this NWP does not authorize any future discharge of dredged or fill material associated with the reversion of the area to its prior condition. In such cases a separate permit would be required for any reversion. For restoration, enhancement, and creation projects conducted under paragraphs (a)(1) and (a)(3), this NWP also authorizes any future discharge of dredged or fill material associated with the reversion of the area to its documented prior condition and use (i.e., prior to the restoration, enhancement, or creation activities) within five years after expiration of a limited term wetland restoration or creation agreement or permit, even if the discharge occurs after this NWP expires. This NWP also authorizes the reversion of wetlands that were restored, enhanced, or created on prior-converted cropland that has not been abandoned, in accordance with a binding agreement between the landowner and NRCS or FWS (even though the restoration, enhancement, or creation activity did not require a Section 404 permit). The five-year reversion limit does not apply to agreements without time limits reached under paragraph (a)(1). The prior condition will be documented in the original agreement or permit, and determination of return to prior conditions will be made by the Federal agency or appropriate State agency executing the agreement or permit. Prior to any reversion activity, the permittee or the appropriate Federal or State agency must notify the District Engineer [COE] and include the documentation of the prior condition. Once an area has reverted back to its prior physical condition, it will be subject to whatever the Corps regulatory requirements will be at that future date. (Sections 10 and 404). (*Federal Register,* Vol. 65, No. 47, March 9, 2000)

FEDERAL REGULATORY APPROVAL

Determining whether a permit is necessary takes time and effort, but it will be a learning experience for all involved. If a project involves digging, dredging, altering hydrology, grading, or placing structures in or around **jurisdictional wetlands**, then check with a regulatory agency to determine whether the project requires a permit.

Jurisdictional wetlands are wetlands that have been identified by the COE as being subject to Federal regulation under the Clean Water Act. If wetlands already exist on the schoolyard, check local wetland maps to find out if the wetlands have already been federally recognized. Wetland maps are typically found at your local planning and zoning office. If the planning and zoning office does not have wetland maps, contact the National Wetland Inventory office (see References) or visit their web site to locate a map with the school property on it. If the wetland maps do not reveal any wetlands on the schoolyard, call the COE. When speaking with the Corps regulators, describe the project. Ask if they will come to the school to perform a jurisdictional delineation, in which a trained regulator determines the outer limits of a wetland area. The COE has jurisdiction over any activities taking place within the wetland. This means that the COE can regulate what activities are allowed to occur in that wetland. The regulator will most likely place colored flags to mark the outer edge of the wetland.

When the COE comes to the site, accompany the regulator to the area of the proposed wetland restoration or enhancement project. Describe the project. Show the regulator any plans that may have been developed. The regulator should be able to provide advice on how to go about the project in a way that will satisfy any regulatory requirements. The regulator should also be able to explain what permits will be necessary for the project. Most likely, only basic information about the project will need to be filed for a general Nationwide Permit allowing restoration, enhancement, or shoreline stabilization activities. In return a Letter of Authorization will be provided by the COE, giving the school permission for the planned wetland project.

The COE offices typically are understaffed, and it may be more expedient to have a local wetland consultant take a quick walk through to see if the school has wetlands on site and if so, to delineate them. The COE then should be asked to verify the delineation. Once verified, the wetland's line should be surveyed and placed on the school's property plat, then sent to the COE for their written approval. This process may take several months.

Table 5 reviews some planned wetland projects and the possible regulatory authorization needed. Be patient and allow a lead time of two to three months to file the permit applications and obtain approval. For more detailed help with the regulatory process, call the regional COE office; describe the project and ask if it will be necessary to file for a permit. Be prepared to make several phone calls, since answers to questions may not be immediately available. Upon finding people who can help, be sure to obtain their names and extensions for future reference or additional information later. While it is best to begin with the COE, also check with state, county, and local authorities to determine if they will require permits as well.

STATE AND LOCAL AUTHORITIES

First, check a local telephone book to locate a county agency for natural resources or for planning and zoning. If appropriate listings cannot be found, check the telephone book for the state agency that handles natural resources and wetlands. For example, Maryland's Department of the Environment (MDE), Virginia's Marine Resources Commission, New York's Department of Environmental Conservation, and New Jersey's Department of Environmental Protection and Energy all handle wetland permitting for their respective states. Some states even have a permit hotline that can be called to find out whether a permit is needed or to check on the status of a permit application.

With access to the internet, an agency that handles wetland permits can be located by searching the state government web site. Some states will provide information on exactly how to apply and whom to contact on the internet. Usually the state regulatory personnel can help figure out what information to submit and how long the permit application will take to process. If unsuccessful in finding a local office that handles wetland permits, state agencies may also provide information on whether a local permit is necessary.

Table 5 Types of Wetland Projects and Possible Regulatory Authorization Needed

Wetland Project	Authorization Needed
Digging out an area within wetlands to create standing water (for a pond).	Individual 404 or Nationwide Permit, local permit, or Letter of Authorization.
Placing erosion control devices along a shoreline (biologs, riprap).	Nationwide Permit or local authorization.
Digging holes to plant wetland plants in a wetland area.	Nationwide Permit, Letter of Authorization, or local permit. Possible Individual 404 Permit needed depending upon size of project. If planting is greater than one-half acre, a permit may be needed.
Building structures such as boardwalks and overlooks in wetlands.	Individual 404 Permit, Nationwide Permit, or Letter of Authorization, depending upon the magnitude of the project.
Grading a sloped area for planting or for a pond (with hydric soils present).	Individual 404 Permit, Nationwide Permit, or Letter of Authorization, depending upon the amount of material being moved and impacts to the wetlands.
Restoring or enhancing a low, >500 ft^2 wet area through planting wetland vegetation.	Possibly no permit. May only need a Letter of Authorization, if anything at all.
Constructing a wetland in a low area using artificial hydrology.	No permit required unless the area qualifies as a wetland. Check to see if the soils are hydric and if the area is listed on a wetland map. Check with a regulator before starting project.

Some municipalities require separate permits, while others rely on state and federal agencies to handle wetland permits. For example, to restore a wetland in the City of Annapolis it would be necessary to obtain a permit from the City Board of Public Works in addition to a permit from the MDE and/or the COE.

If neither the local nor the state government handles wetland permits in your region, then contact the regional COE office to determine whether a Nationwide Permit under the Clean Water Act will be necessary.

FEES

Most regulatory agencies charge a fee for permit processing. Fees can range anywhere from a nominal amount ($0 to $15) to a significant expense ($100 to $5,000). Typical schoolyard wetland projects will not require extensive permits; therefore, the permit fee will probably be low.

PERMIT APPLICATIONS

Some important information will need to be collected to complete the permit application. While the information requested varies, this section provides general guidance regarding the most common types of information needed.

Three different types of project drawings may be needed: a **vicinity map**, a **plan view**, and an **elevation** and/or **cross section view**.

The Vicinity Map

The Vicinity Map can be a road map or a U.S. Geological Survey topographical map. Copies of large-scale, detailed maps are usually available at the local planning and zoning office. The school may have a map showing the property lines and the surrounding area where the project will take place. The map should have enough detail to show interested parties exactly where the project site is located. Draw an arrow pointing to the exact location of the site. Be sure the map includes names of waterways, existing landmarks (e.g. the school), the name of the community, names of roads, a directional orientation (north arrow will do), and a scale.

The vicinity map is the map that would be printed and distributed to the public if a public notice is required for the project. Depending on the size and type of project, the government may or may not require a public notice to be sent. Usually a public notice is reserved for situations where there will be significant negative impacts to wetlands.

The Plan View

The Plan View is a pictorial description of a project from a bird's-eye view. The Plan View should include surrounding waterbodies (rivers, creeks, lakes), existing shorelines, mean high and mean low water lines, average water depths around the project, the dimensions of the activity and proximity to waterways or wetlands, location of adjacent structures (e.g. the school, maintenance buildings), location of wetlands, scale, and north arrow. Mean high water information should be available from the local planning and zoning office. Students can draw the plan view based on the information they collected to develop the project. For wetland restorations, include planting areas, any area that will be cleared or excavated, and any structures that will be placed on the site (e.g. boardwalk, fence).

The Elevation or Cross Section View

The Elevation or Cross Section View will show a side view of the project's elevations. This drawing will show water elevations, water depth, mean high water line, a cross-sectional description of any excavation, approximate slopes (based on elevation measurements), and scale. Be sure to include all major dimensions of the project. For example, if part of an existing wetland will be excavated to create a small ponded area, the Cross Section View should show the depth of excavation. The Plan View will show the area of excavation. **Figures 3.1** and **3.2** on the following pages are examples of permit application drawings. The drawings need not be highly refined, but they must be understandable.

The Project Description

This is the most important section of the permit application. Explain the purpose of the project (e.g. to restore/create a wetland for educational and research use and to show students how to restore/create a wetland). Describe exactly what will be done (excerpt this from the plans and specifications). Give the range of dates between which the project is expected to start and be completed. Give the project's dimensions. For example, if restoring a wetland area along a shoreline, state the number of feet of shoreline that will be restored. Give the square footage of area that will be planted. Describe any nonnatural material that will be used on the site (for example, biologs for erosion control, or stone to hold or absorb wave energy to allow a marsh to develop). Much of the above information will be in the drawings that are submitted with the application. Include with the application written authorization by the school authorities (the school principal, or someone from the county board of education).

SCHOOL AUTHORITIES

Some school systems require teachers to place fences or install other safety features around wetlands and open water. To open the lines of communication about safety and maintenance issues, hold a meeting with the school principal, the maintenance supervisor, and the teachers who will be using the wetland. Share the plan for making the wetland safe and for maintaining the wetland. Keep lines of communication open with the principal and any local-level school authorities to make sure that everyone fully supports the project.

REFERENCES

National Wetlands Inventory, 9720 Executive Center Drive, Suite 101 Monroe Building, St. Petersburg, FL 33702. Telephone: 813-570-5412; fax: 813-570-5420. Website: www.nwi.fws.gov

U.S. EPA Wetland Information Hotline: 800-832-7828

U.S. Army Corps of Engineers website: www.usace.mil/usace-hq.html (for regulatory offices and permit applications)

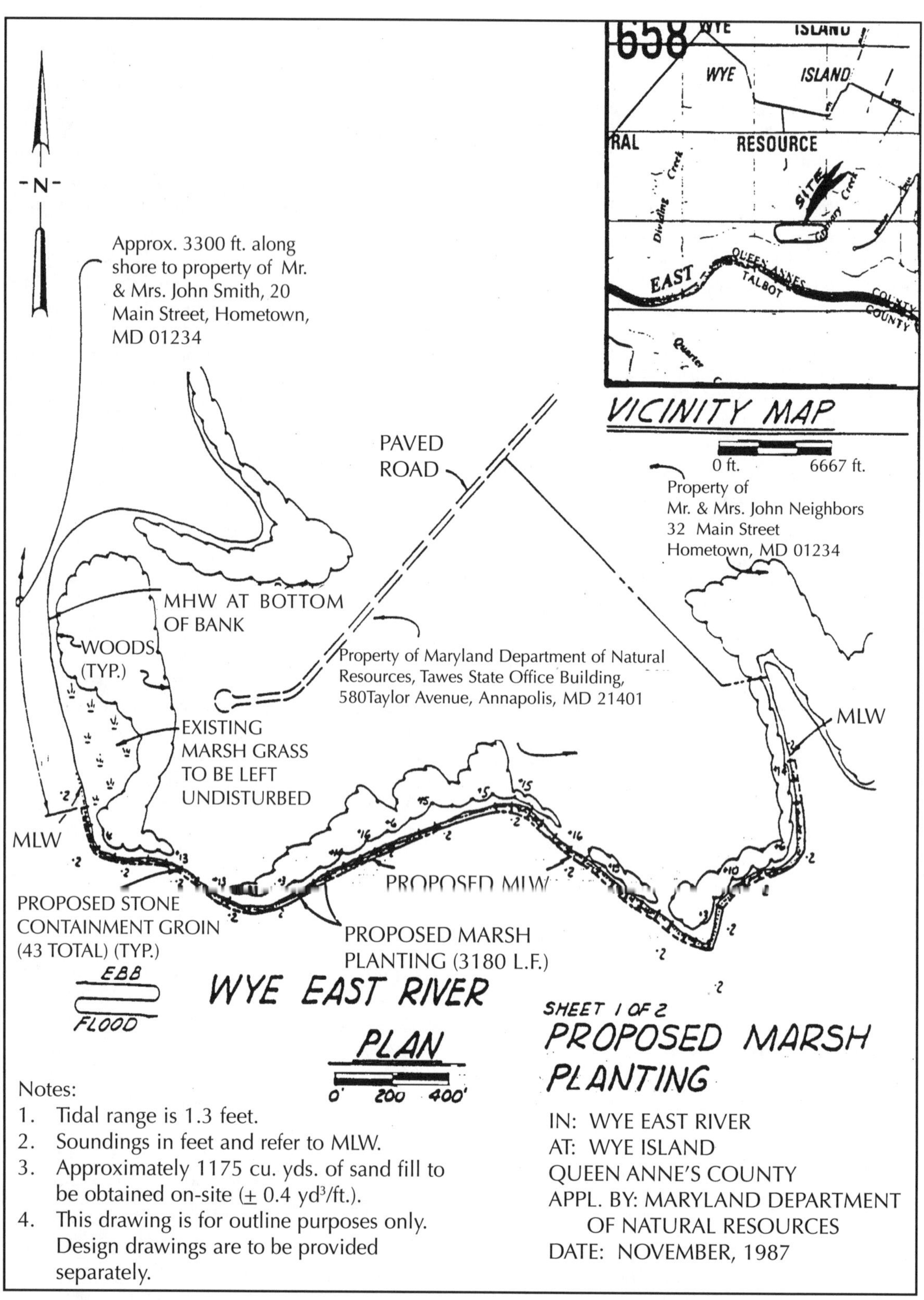

Figure 3.1 A Plan View and a Vicinity Map of a proposed shoreline restoration along the Wye East River.

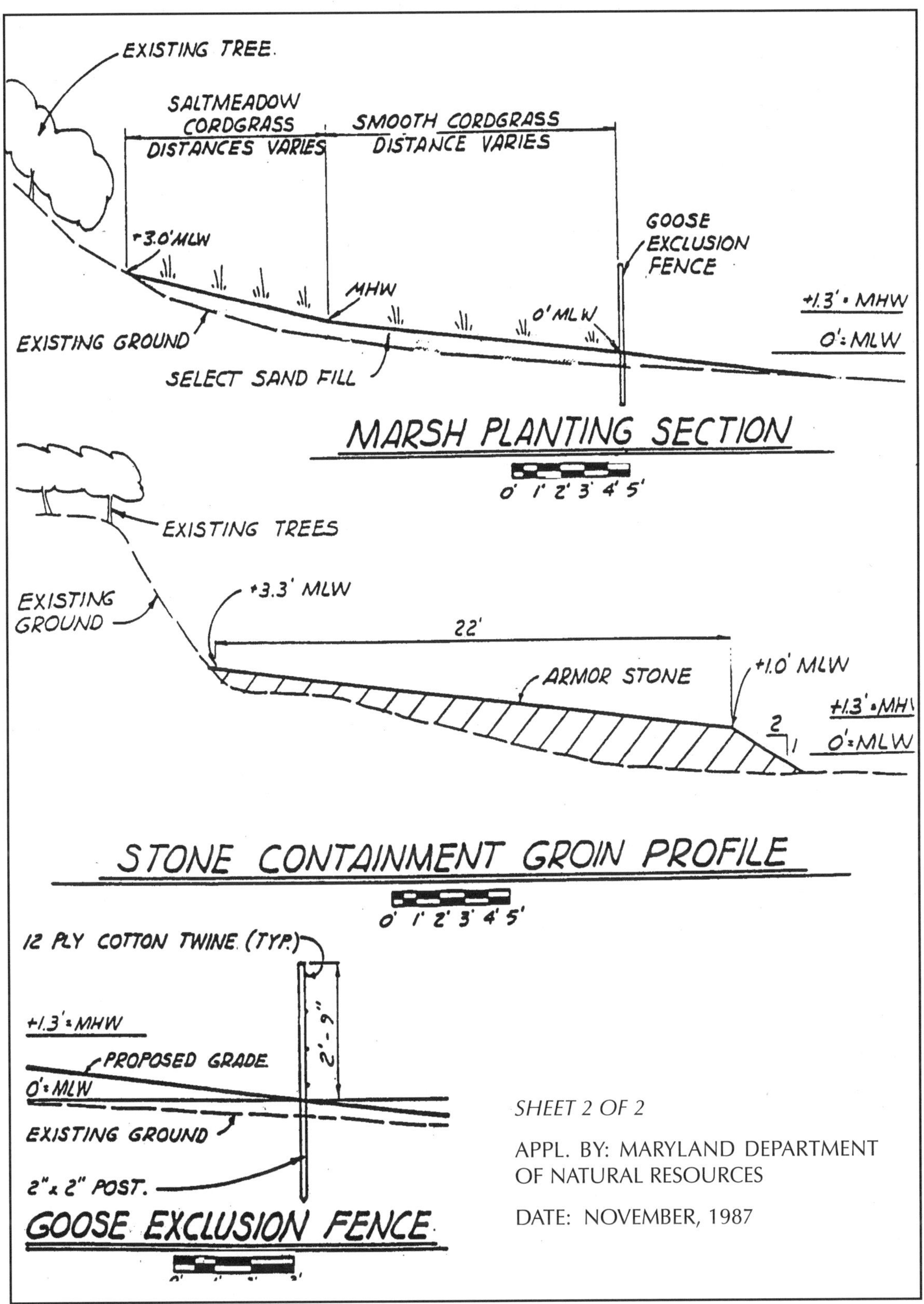

Figure 3.2 Cross Section Views of the shoreline restoration shown in Figure 3.1.

4 DESIGN, GRADING PLAN, AND SPECIFICATIONS

Prior to constructing, restoring, or enhancing the wetland, specifications for the earthwork and planting must be developed. In order to develop a grading plan for the planned wetland, the following are needed: (1) its design and (2) one-foot contour lines for the area where the planned wetland will be placed. If these contour lines are available, go directly to page 58 to start work on the wetland design. If these contour lines are not available, they may be acquired by hiring a local surveyor to do the necessary topographic surveying. Alternatively, educators and students can do this topographic surveying.

ELEMENTARY TOPOGRAPHIC SURVEYING

To complete the necessary topographic surveying to obtain contour lines for the planned wetland site, you need the following equipment and supplies:

- a level or a transit with tripod,
- a pocket compass,
- an elevation rod made of wood or plastic (generally marked off in tenths of feet with further gradations between the tenth marks and generally extendable to 32 feet or ten meters),
- a measuring wheel or tape measure,
- several two inch by two inch wooden stakes two to three feet long, several heavy nails two to three inches long, colored surveyors' flags, and bright plastic tape.

Surveying the Site

Step 1. Establish a project benchmark within 50 feet to 300 feet of the actual construction activities of the planned wetland. The benchmark should be considered permanent, be located where it will not be damaged or its elevation altered, and be on a level surface. The benchmark can be a fire hydrant, a level concrete slab, an existing surveyor's benchmark, a nail in a post, or a flat-topped two inch by two inch wooden stake hammered to about six inches above the ground surface. Once the benchmark has been established, determine its elevation if it has been used as a benchmark in the past. If it is a new benchmark, assign it an arbitrary elevation. Keep in mind that when measuring elevations downhill to X feet, an elevation that is > X feet should be assigned so that all topographic elevations will be positive.

Step 2. Set up the level or transit on a tripod that is firmly secured to the ground. Its location and height should allow easy reading of the elevation rod when placed on the benchmark and on as many of the other topographic points as possible without having to relocate the tripod and instrument. If the instrument is not self leveling, level it using the leveling screws. Continue leveling, so that upon rotating the instrument on the tripod by 360°, the level bubble remains stationary.

Step 3. Establish a linear baseline from which perpendicular transects will be established at its beginning, at its end, and at regular intervals (X) between the two. The transects are to pass through the area to be graded. The baseline length should be a multiple of X. For example. If X = 25 feet, a baseline length may be 400 feet, but not 415 feet. If X = 25 feet, insert a flag at the beginning and end of the baseline and at every 25 feet in between. The flags represent the starting point of each transect and are assigned station numbers. For X = 25 feet, the 1st station is Station 0 + 00, the 2nd station is Station 0 + 25, the 4th station is Station 0 + 75, and so on.

Step 4. Using the pocket compass at Station 0 + 00, establish the magnetic bearing of the baseline by sighting along the station flags that have just been set. Turn 90° (right angle) to the baseline by adding to (or subtracting from, as appropriate) the compass setting. While sighting along the 1st transect at 90° to the baseline, have a second person place several flags along the transect line. Repeat the above at each station.

Step 5. Have the rod person place the rod vertically on the project benchmark. The instrument person then sharpens the cross hairs and focuses the instrument on the rod. The Height of Instrument (HI) then is recorded as the Rod Reading (RR) plus the assigned Benchmark Elevation (BE) using **Equation 4.1**.

$$HI = RR + BE \qquad \textbf{(4.1)}$$

The rod person with the measuring wheel then goes to station 0 + 00 and sets the rod vertically while the instrument person records the RR. The rod person then proceeds along the Station 0 + 00 transect with the measuring wheel stopping every Y feet (usually 25 feet) so the instrument person can record the RR. The Ground Elevation (GE) for every RR is given by **Equation 4.2**.

$$GE = HI - RR \qquad \textbf{(4.2)}$$

Repeat the above along each station transect until all the desired elevation points are obtained.

Do not worry if it takes more than one day to complete a topographic survey. At the end of the day dismantle the instrument and return it, the tripod, and the rod to the classroom or office. When setting up the instrument another time, the HI will be different from the first setup; however, using the same project benchmark will ensure that all the GEs have not changed.

Moving the Survey Instrument

When unable to obtain all necessary elevation points from the initial instrument setup, move the instrument to a new location to complete the topographic survey. To do this while keeping all of the ground elevations referenced to the original benchmark: (1) Obtain the GE_{TP} at an intermediate point, called the Turning Point (TP), between where the instrument was first located and where it will be moved. (2) Put a flag in the ground at this TP. Set up the instrument at the new location and obtain a RR_{TP} at this TP. The new Height of Instrument (HI_{NEW}) is equal to RR_{TP} plus the GE_{TP}. Continue getting GE data, all relative to the original project benchmark, using **Equation 4.3**.

$$GE = HI_{NEW} - RR \qquad \textbf{(4.3)}$$

See **Figure 4.1** for a review of the process of moving the instrument.

With the instrument at the new location, a new baseline and station transects will most likely need to be established in order to complete the topographic survey.

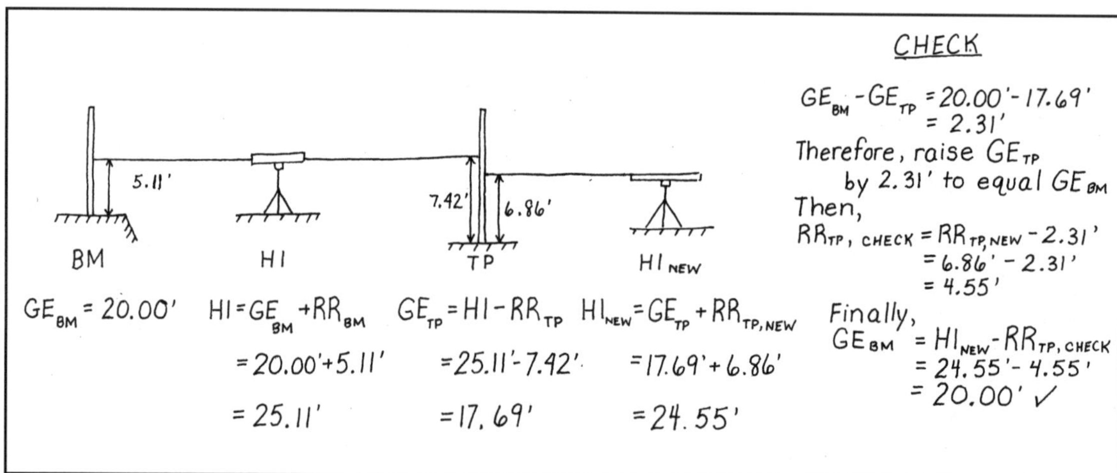

Figure 4.1 Sample calculations for moving the survey instrument to a new location for completion of a topographic survey.

DRAWING A TOPOGRAPHIC MAP

To draw a topographic map, first plot to scale all the elevation points that were surveyed. Then draw contour lines throughout the plotted data at 1.00 foot intervals. The resultant plot of contour lines is the topographic map. An example of a topographic map is given in **Figure 4.2**. This data will be used to illustrate the process of constructing a planned wetland.

Figure 4.2 is a topographic map, with data points included, of part of a schoolyard where stormwater runs down a shallow swale (contour lines pointing uphill define a swale) and is discharged over a steep embankment to a tributary of the Chesapeake Bay. The watershed for the swale is seventy-five times larger than the area shown in **Figure 4.2**. Consequently, normal precipitation events lead to significant volumes of stormwater flowing through the swale. Shown in **Figure 4.2** are (1) a concrete slab which was used as the project benchmark and assigned an elevation of 20.00 feet, (2) the baseline with seven station transects at twenty feet intervals, (3) the topographic data points, (4) the contour lines at 1.00 foot intervals, and (5) the top of the bank line, after which the elevations drop off rapidly to a tributary of the Chesapeake Bay.

Topographic maps showing closely spaced contour lines indicate that such areas have steep slopes or grades. Areas having shallow or nearly flat slopes or grades are represented by widely spaced contour lines. A series of contour lines whose arcs point downhill represents a ridge. A series of contour lines whose arcs point uphill represents a swale.

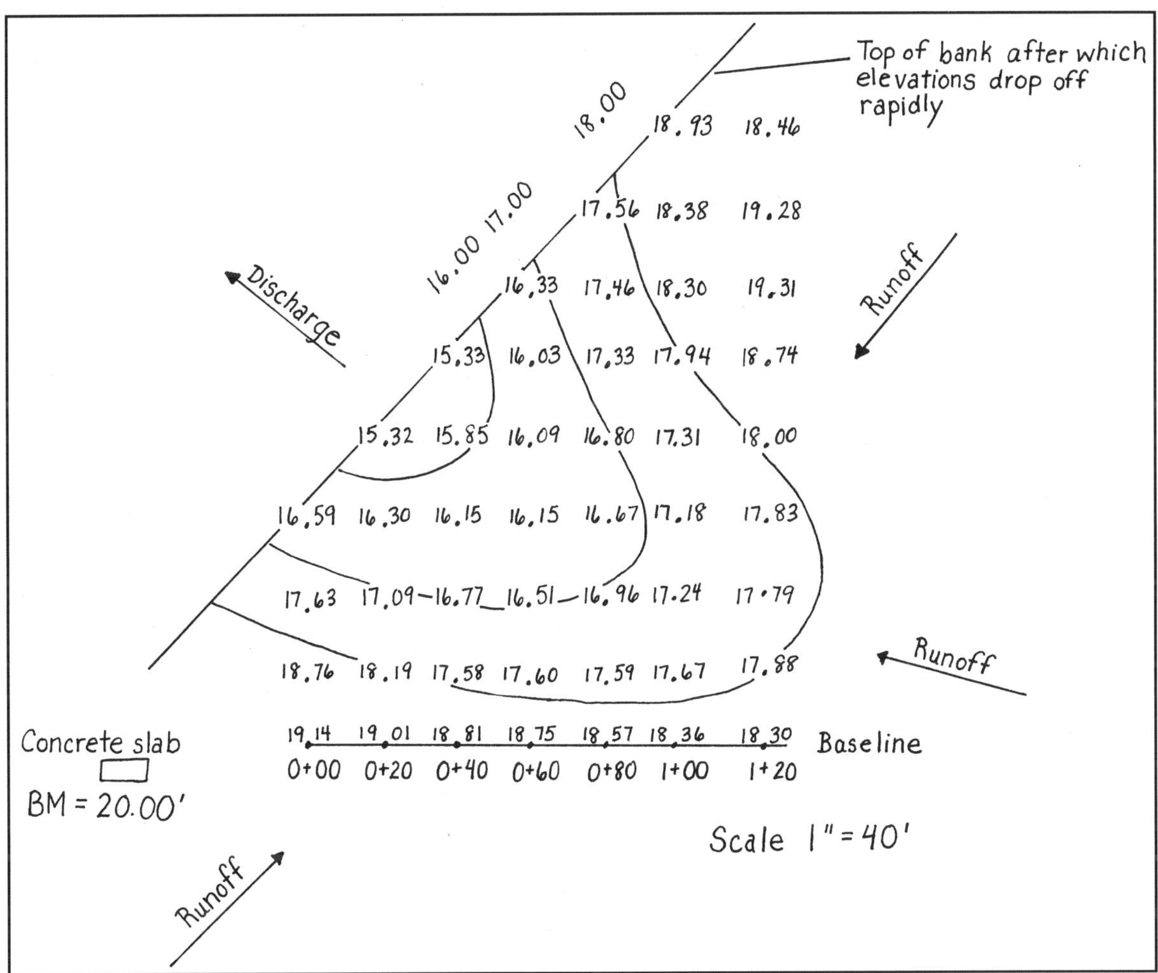

Figure 4.2 A topographic map of a schoolyard site for a planned wetland.

DESIGNING A PLANNED WETLAND

The grading plan for a planned wetland is a graphic representation of what the topography of the site will be after the planned wetland is constructed. Consequently, before a grading plan can be prepared, the planned wetland must be designed.

As no wetland previously existed or presently exists in the area shown in **Figure 4.2**, the planned wetland will be a constructed wetland.

Objectives for a Planned Wetland

The objectives of a planned wetland must be clearly defined in the design phase of the project, as these objectives will influence the planned wetland's hydroperiods and depths of flooding, source of water, placement, size, and various structural elements of design. If the objective of the planned wetland is to provide the opportunity to research the effectiveness of the wetland to improve water quality, its design will be different from those having the objective of providing aesthetics to the schoolyard, nesting opportunities for various bird species, habitat for various wildlife species, or the greatest possible diversity of plant species for botany classes.

As the objectives of the wetland have to meet as many of the interests of the teachers as possible, teacher meetings should be held to discuss these interests. Many of the interests cannot be satisfied by a single wetland design, as shown above. However, it must be determined if the conditions are present to design a single wetland that will satisfy the maximum number of these interests. If the hydrology or the potential size of the planned wetland is insufficient to achieve these objectives, the design must be adjusted to achieve as many of these objectives as possible.

An Example of Objectives

Suppose that the objectives of a permanently flooded marsh are to research the differences in invertebrate populations, in vegetation dynamics (natural changes in vegetation communities), and in bird utilization between the planned permanently flooded marsh and an irregularly flooded wet meadow that occupies another area of the schoolyard. Both wetlands are in areas subjected to full sun. The area shown in **Figure 4.2** is about 0.3 acres, and its watershed is 23 acres. One might expect that the volume of stormwater from normal precipitation events would be sufficient to maintain a permanently flooded marsh. However, before proceeding with a design and grading plan, hydrology verification of the desired marsh design is necessary to determine if the desired design will work or if the design must be modified to achieve the required hydrology.

Hydrology Verification of the Design

If the planned wetland has an unlimited or abundant water supply and the supply is not pulsed, as from a stream where water can only be taken above its base flow, hydrology verification is not needed. In such instances it is important that biological benchmarks be available to set the final grades for the desired plant communities.

When it is not known whether the water supply (groundwater springs, stream water above base flow, stormwater runoff) is sufficient for the planned wetland's needs, the desired design must be subjected to a hydrology verification. Knowing the hydroperiods and the depths of flooding of the desired planned wetland design, perform water budget calculations to determine whether the desired design will be successful hydrologically or if design adjustments (size, water depths, hydroperiods) must be made. Perform water budget calculations on all design adjustments until a final design is reached whose hydrology is acceptable.

An Example of Hydrology Verification

Since the planned wetland will be supported only by stormwater runoff, Chapter 2 should be reviewed. First calculate R_{in} to the planned wetland area in **Figure 4.2**. (See **Equation 2.5**.) This runoff discharges at the discharge point, which is located at the beginning of the arrow which is labeled "discharge" in **Figure 4.2**.

Of the 23.3 acres of drainage area, 0.3 acres (1.3%) will contain a permanently flooded marsh, most of which will be vegetated; 3.5 acres (15.0%) will be covered by impervious roofs and pavements; and 19.5 acres (83.7%) will contain turf with >75% grass cover. The soils are silty loam throughout the entire 23.3 acres, including at all wetland grades, assumed to be from one foot to three feet below the existing grades. The hydrologic soil group for the watershed is Group B (**Table 3**, page 41). With the above information the applicable runoff coefficient is determined:

$$c = (0.013)(0.84) + (0.15)(0.94) + (0.837)(0.18) = 0.303$$

Suppose that the project site is near Easton, Maryland where the mean monthly temperature and precipitation during the growing season are 75° and 4.4 inches respectively. Because the mean two-week precipitation is 2.2 inches, which is greater than two inches but less than three inches, the applicable runoff coefficient is reduced by 20% and becomes:

$$c = 0.303 - (0.2)(0.303) = 0.242$$

Assuming that the planned wetland will contain approximately 25% of its area in open water 3.5 feet deep and 75% of its area in marsh mostly covered by 0.5 foot of water, the entire 0.3 acre wetland will have an average water depth of 1.25 feet at its pool level. Therefore, to fill the empty wetland will require (0.3 acre)(43,560 ft^2/acre)(1.25 ft) = 16,335 ft^3 of runoff.

From **Equation 2.5**, the R_{in} for the average two-week precipitation during the growing season is:

$$R_{in} = (23.3 \text{ acre})(0.242)(2.2 \text{ inch})(3{,}630 \text{ ft}^3/\text{acre inch}) = 45{,}030 \text{ ft}^3$$

A substantially greater water supply is desirable for filling a dry wetland during the average two-week precipitation. Consequently, ΔS in Equation 2.4 should be $\leq 0.5\ R_{in}$. Using the volumes of runoff obtained every two weeks and that are required to fill the empty wetland, ΔS in **Equation 2.4** = 16,335 ft^3. Since this value is $\leq 0.5\ R_{in}$ a wetland of approximately 0.3 acres or smaller will receive every two weeks more water than necessary to fill the wetland.

The following question now must be answered. Is the rate of water loss from the wetland sufficiently low so that the wetland will be permanently flooded? If in every two weeks there will be more than sufficient precipitation to fill the empty wetland, will the wetland, primarily at the 0.5 foot water depth, remain flooded during a two-week period?

Considering the wetland just after it fills to its pool level, what will be the water loss before the next two-week precipitation? **Equation 2.3** simplifies to **Equation 2.6** which gives the water loss due to infiltration (I) and evapotranspiration (ET). These water losses are the same as given in the earlier example on page 43, since the average monthly summer temperatures of 75°, site locations (latitude of 40°), and the soil types (silt loam) are the same. ET is calculated to be -0.02 ft/day and I is calculated to be -0.54 ft/day. Consequently, the water loss is 0.56 ft/day and the marsh which is covered by 0.5 foot of water at its pool level, will go dry one day after a two-week precipitation event. Clearly, it is impossible for the marsh to be permanently flooded. The marsh will have to be treated by one of the four methods outlined on page 44 in order to greatly reduce or eliminate the loss of water due to infiltration. Under these conditions, the water loss will be approximately 0.28 foot during a two-week period.

Final Design of the Planned Wetland

Since the hydrology verification of either the desired planned wetland design or an adjustment of this design was confirmed, it is now appropriate to summarize all of the design elements here.

An Example of a Final Design

Since hydrology verification of the desired marsh design was confirmed, provided the loss of water due to infiltration is essentially eliminated, the final design for this marsh is as follows:

1. The total wetland area must be 0.3 acres or less.
2. The marsh area permanently covered by mostly 0.5 foot of water and then sloping up the edge of water and down to 1.0 foot of water must be 0.23 acres or less.
   a. The open water area having 3.5 feet of water depth must be 0.07 acres or less.
   b. The marsh comprises 75% or more of the total wetland.
   c. A small island in the open water area provides nesting and resting opportunities for birds.

GRADING PLANS

The grading plan for the planned wetland is a graphic representation of what the topography of the site will be after the planned wetland is constructed. A grading plan is needed if any earthwork (excavation or filling) is required to complete the project.

To start the grading plan, pencil in the new contour lines for the planned wetland on a copy of the topographic map. Then copy (trace) these lines onto a clean piece of paper. This will be the grading plan. Draw one or more lines through the plan. These lines will show the elevation changes along their lengths, and are called cross sections or sections.

It is important to remember that wherever the new topographic lines are placed, their ends, if open-ended and not closed (as for a circle or parallelogram), must be connected to the same elevation somewhere on the plan.

An Example of a Grading Plan

In order to create a 0.3 acre or less permanently flooded marsh in the area shown in **Figure 4.2**, the earthwork will include excavation throughout the 16 feet to 18 feet contour lines, and filling along the top of bank to create a berm and a water control discharge area. The design is to create up to 0.23 acres of marsh from the water line (pool level) down to a water depth of 1 foot, including a major area, a permanently flooded marsh shelf that will be flooded by 0.5 foot of water at pool level. Below a water depth of 1 foot, the area is further graded to an open water area of up to 0.07 acres that has a maximum water depth of 3.5 feet at pool level. In the center of the open water area a small upland island will be constructed by filling to provide a relatively protected environment for bird nesting and resting.

The grading plan for the marsh together with Section A-A is shown in **Figure 4.3**. In this figure, contour line 18.00 is the only one that is open ended at both ends of the berm. This contour line must connect with the existing contour line, as shown.

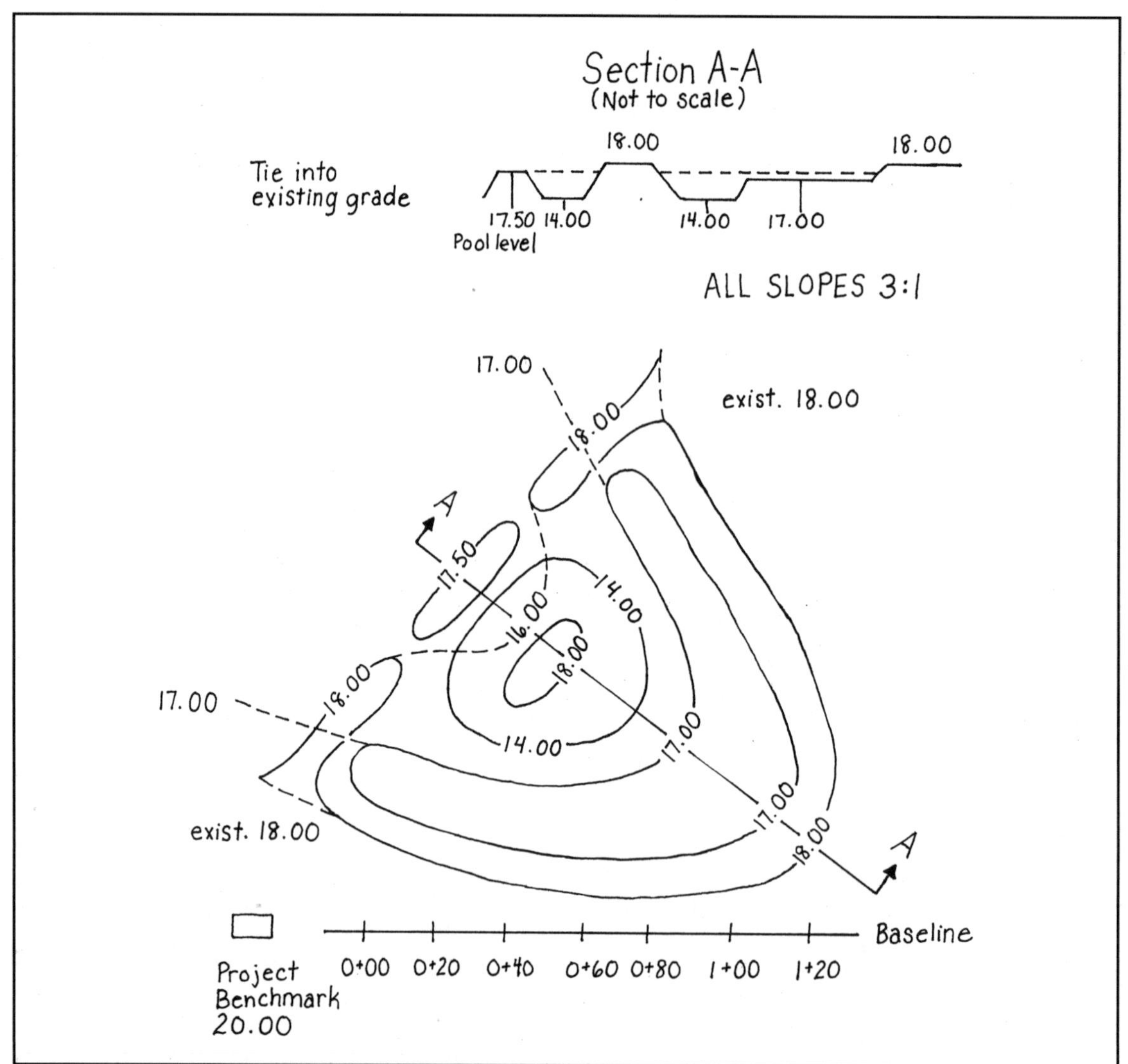

Figure 4.3 An example of a grading plan for a permanently flooded marsh.

Shown in this grading plan are:

- a total wetland area of 0.21 acres;
- 0.16 acres of marsh from elevation 17.5 (pool level) to elevation 16.5;
- 0.11 acres of marsh shelf at elevation 17.00;
- 0.05 acres of open water from elevation 16.5 to elevation 14.00;
- 0.01 acres of upland island;
- a permanent pool level of 17.50;
- the project benchmark at elevation 20.00; and
- the baseline.

PLANS AND SPECIFICATIONS

The plans and specifications for the planned wetland serve as the construction documents. They must be sufficiently accurate and complete so that a general contractor and a landscape contractor can use them to construct a successful planned wetland if constructed according to them. The plans and specifications will serve as a permanent record of the project. They will also serve as the basis for future monitoring of the planned wetland, as they will accurately reflect the objectives of the project.

The grading plans should clearly show at the least:

- all of the applicable water levels;
- bottom elevations of all water control structures, both intake and discharge;
- the locations of all physical benchmarks with corresponding elevations;
- sections showing all conditions in the planned wetland;
- 1.0 foot contour lines; and
- access points, structures and buildings, and property lines, if important.

The grading specifications should clearly indicate at least:

- quantities and specifications for all listed materials (e.g., stone, added topsoil, geotextile materials);
- the locations for stockpiling or disposal of excess excavated materials;
- a construction timetable that spells out necessary sequences, such as the all important seeding timetable;
- special considerations and conditions when critical for project success, such as use of mats and low ground pressure equipment;
- details for substrate deconsolidation in areas to be planted through rototilling, discing, or ripping; and
- that the construction contractor be responsible for flagging the various planting zones.

The planting specifications should clearly provide at least:

- a list or table of the plant species, types of plant materials, and plant quantities for planting;
- a list or table of the plant species with percentage composition and weights for seeding;
- a list or table of acceptable substitutes;
- planting details, including plant spacing, fertilization, and soil amendments;
- seeding details, including any specialized techniques;
- watering requirements, if important; and
- details of any required wildlife control structures.

When developing the above, it is useful to have equations to calculate quantities of plants, fertilizer, and soil amendments. **Equation 4.4** can be used to calculate the quantities of planting units once the plant spacing (on-center spacing) is specified.

$$\text{Number of planting units} = \frac{(\text{area in ft}^2)}{(\text{on-center spacing in ft})^2} \qquad \textbf{(4.4)}$$

Equation 4.5 can be used to calculate the pounds of Osmocote® fertilizer that are needed to fertilize the herbaceous plants with 30 grams (one fluid ounce) of fertilizer per plant.

$$\text{Pounds of fertilizer required} = \frac{(\text{number of planting units})(30 \text{ grams/unit})}{(454 \text{ grams/pound})} \qquad \textbf{(4.5)}$$

This fertilizer is available in the following three useful time frame releases:

- For summer planting, use three- to four-month release.
- For spring planting, use eight- to nine-month release.
- For winter/fall planting, use twelve- to fourteen-month release.

When planting woody plants, it is recommended to use Agriform® two-year release, 10-gram tablets at the manufacturer's recommended rate. The recommended soil amendment when planting woody plants is leaf or pine bark compost. Equal parts of soil and pine bark or leaf compost should be used with the fertilizer to backfill the planting hole. The leftover soils from creating the planting holes should be raked about the plants so as not to alter the surrounding grades significantly.

Assuming that the planting hole is approximately twice the size of the container, **Equation 4.6** may be used to estimate the required amount of compost.

(# of plants)(container size in gal)(0.005 cu yds/gal)
= cu yds of compost **(4.6)**

For 5,000 plants in quart containers, the following amount of compost will be needed:

(5,000)(1 qt.)(1 gal./4 qt.)(0.005 cu.yds./gal.)
= 6.25 cu. yds of compost

An Example of Grading Specifications

1. Stockpile the subsoils and topsoils separately on site, but not where surface water runoff will be blocked. Use these soils to construct (1) the berm at elevation 18.00 and the discharge spillway at elevation 17.50, and (2) the island. Spread any excess materials throughout the drainage area adjacent to the wetland. Should additional fill materials be required, obtain them by deepening the open water area.
2. The shelf area at elevation 17.00 is not to be topsoiled. Topsoil only the berm, spillway, and island.
3. Should precipitation occur before the grading is complete and stormwater becomes ponded, pump the water to the discharge point before beginning the grading.
4. Place flags along elevation 16.00 just prior to completion. This is the low elevation for the herbaceous plants.
5. Rototill the shelf area at elevation 17.00 to a depth of three inches just prior to completion. This will deconsolidate the soils that are to be planted with herbaceous plants.
6. Contact the landscape contractor as soon as construction of the discharge spillway is complete, so this area can be seeded and stabilized as soon as possible.

An Example of Planting Specifications

1. See **Table 6** for details of the herbaceous aquatic plants and **Table 7** for details of the woody plants for the island.

2. All areas at and above elevation 17.50 that have been disturbed shall be seeded with the following seed mixture:

chewings fescue (*Festuca rubra)*	50%
annual ryegrass (*Lolium multiflorum*)	40%
white clover (*Trifolium repens*)	6%
red top (*Agrostis alba*)	4%

Seed at any time of the year and at a rate of 80 lbs/acre.

3. Acceptable substitutes for the herbaceous aquatic plants are:

Scirpus fluviatilis for *Scirpus pungens*
Saururus cernuus for *Pontederia cordata*
Sagittaria latifolia for *Peltandra virginica*

An acceptable substitute for the woody plants is:

Ilex verticillata for any of the shrubs specified.

4. All herbaceous plants shall be planted in the dry (water pumped out of the wetland, if necessary, to provide no more than one inch of water covering the planting zones).

Planting may be accomplished at any time of year that the ground is not frozen.

Planting holes shall be prepared using a mechanical auger or a shovel that is at least twice the width and the same depth as the plant material.

One fluid ounce (30 grams) of the appropriate time release Osmocote® fertilizer shall be placed in the bottom of each planting hole. A total of 30 pounds of fertilizer will be needed.

Peat potted plants shall be placed directly into the planting holes and plants in quart containers shall be removed from the containers and then placed into the planting holes. All plant material shall be placed on top of the fertilizer. The heights

Table 6 Herbaceous Aquatic Plants

Elevation Range (Area)	Species	On-center Spacing	Type of Plant Material	Unit Quantity
17.00 to 17.50 (1,890 ft^2)	*Scirpus pungens*	2 feet	1.75 inch peat pot	473 each
17.00 (4,600 ft^2)	*Pontederia cordata*	3 feet	1 quart pot	511 each
16.50 to 17.00 (1,350 ft^2)	*Peltandra virginica*	3 feet	1 quart pot	150 each

Table 7 Woody Plants for the Island

Species	On-center Spacing	Type of Plant Material	Unit Quantity
Cephalanthus occidentalis	6 feet	1 gallon container 2 to 3 feet tall	4 each
Clethra alnifolia	6 feet	1 gallon container 2 to 3 feet tall	4 each
Cornus sericea	6 feet	1 gallon container 2 to 3 feet tall	4 each

of root balls shall be adjusted so that their tops range from the ground surface to one inch below. Soil is then packed back into the planting holes to secure the plant material.

5. All woody plant material shall be planted on the island at any time of the year that the ground is not frozen.

Using an auger or a shovel, planting holes shall be prepared that are about twice the diameter of the containers.

The plants shall be removed from the containers and placed into the planting holes.

The heights of root balls shall be adjusted, so that their tops are even with the ground surface or up to one inch below.

A mixture containing equal portions of leaf or pine bark compost and existing substrate is then packed into the planting hole to secure the plant material. A total of 0.60 cubic yards of compost is required. Agriform® time release fertilizer tablets then shall be placed into the soil at the manufacturers recommend rate.

Each planted shrub shall be watered to the point of soil saturation.

6. The specified seed mixture shall be broadcast over all areas at and above elevation 17.50 that are to be seeded. By hand raking or by the use of a tractor with cultipactor on accessible areas, seed shall be brought into maximum contact with the soil. So as not to overseed the shrubs on the island, use a drop seeder rather than a broadcast seeder, to seed around the shrubs.

The discharge spillway at elevation 17.50 shall be seeded as soon after its preparation as possible and immediately covered and anchored with erosion control netting that has been accepted by the project leader.

7. A rain gauge shall be placed on a stake in an open area at the wetland site. The gauge shall be checked weekly during the growing season. All of the shrubs on the island shall be watered with six gallons of water each week during the growing season when there has been less than 0.2 inch of precipitation.

8. The open water area is too small to be used by Canada geese. Consequently, it is not expected that Canada geese will present a wildlife problem. If any wildlife problems arise, wildlife experts should be consulted to determine the best solution.

Common snipe,
Gallinago gallinago

5 CONSTRUCTION AND MAINTENANCE

CONSTRUCTION

The construction of a planned wetland involves many steps. The designed hydrology must be verified. If earth moving is involved, the required sediment and erosion control plan must be implemented. Grade stakes must be installed throughout the site, and the area must be graded according to the plans and specifications. The site must then be planted according to those plans and specifications.

Before starting construction of the planned wetland make certain that:

- all of the necessary federal, state, and local permits are in hand (remember that a local sediment and erosion control plan and grading permit may be needed);
- there is no permit condition that limits the timing of the construction;
- all of the necessary labor, plant materials, supplies, and equipment are or will be on site when needed; and
- the grade stakes have been placed throughout the site. A grade stake is a wooden stake that has marked on it clearly the elevation of fill (for a filling operation) or the depth of cut (for an excavation operation).

Grading the Planned Wetland

Once the necessary surveying to produce the site's topographic map has been done (Chapter 4), install the grade stakes. Check the planned wetland's elevations during grading. If necessary, retain a registered surveyor to complete these tasks.

Grade stakes for the permanently-flooded marsh example discussed in Chapter 4 should be placed along all the final grades. If all the cuts to get from existing grades to the final grade are the same, this means that the existing grades are the same and only a few grade stakes need to be placed. The greater the number of different cuts to get from existing grades to the final grade, the greater the number of grade stakes that will need to be placed. Referring to the grading plan shown in **Figure 4.3** (page 60), place grade stakes, showing the amount of fill or cut, along the following final elevations: 18.00 (start of cut), 17.00 (marsh shelf), 14.00 (open water), 18.00 (island), 18.00 (berm); and 17.50 (discharge spillway). All grades between these elevations have a 3:1 slope.

Providing the designed hydrology is correct, the final grades of the planned wetland then will be the key to success. Consequently, the site's elevations should be continuously checked during the grading phase to ensure that they conform to the grading plans and specifications.

All other components of the grading specifications should be checked by the project leader or a representative of the design team to ensure that they are being performed correctly.

Planting the Planned Wetland

Planting should be initiated following the grading of the planned wetland. In some instances where seeding of wetland plants has been specified to be completed within a narrow seeding window, only that portion of the planned wetland that is to be seeded will be graded first. The balance of the planned wetland will then be graded and planted.

All planting operations must be conducted at times, under conditions, and in manners that are detailed in the planting specifications. A member of the design team or the project leader should be present during most of the planting operations to ensure that the plant materials are acceptable and that the planting specifications are being followed.

If volunteer workers are used during the planting operation, they must be closely supervised and made familiar with the planting specifications. Planting crews no larger than four persons are recommended, and each crew must have its own supervisor, who may be a member of the crew.

MAINTENANCE

When planning a wetland that has a maximum value for waterfowl utilization, alterations in seasonal water level and in associated plantings throughout the wetland may be scheduled. In these instances, the details of such scheduled maintenance should be placed in the plans and specifications. Other maintenance issues will be identified as a result of regular site inspections.

Site inspections should be scheduled monthly and need only constitute a walk through the planned wetland. During such inspections it should be determined:

- if any of the plants require watering due to the amount of precipitation collected in the rain gauge during the previous month;
- if any replanting is required due to plant mortality or plant removal by ice;
- if the water control/conveyance structures (weirs, culverts, channels, etc.) are clogged with debris and need to be cleaned out;
- if the water levels in the planned wetland have been altered by beaver dam construction or muskrat tunneling and need to be restored;
- if there is sufficient evidence of herbivory to warrant constructing exclosure devices;
- if the removal (physically or by herbicide) of undesirable volunteer plants is appropriate; and
- if the removal of deposited wrack (accumulation of leaves, herbaceous stems, and woody debris along the shore of the wetland) is appropriate because of the smothering of herbaceous plants. Generally this wrack is a natural habitat for a host of invertebrates. Even though such wrack may smother plants and be unsightly, it functions to diversify the wetland ecosystem and consideration should be given to not removing it.

Following each inspection, the findings and recommendations should be written up in a maintenance log. When any recommended maintenance repair or activity is completed, it should be so entered in the maintenance log.

Common reed, *Phragmites australis*

6 COST ESTIMATES

The costs for implementation of the planned wetland generally can be divided into two phases:

- the design phase
- the construction phase

Since one cannot estimate the cost of the construction phase until after the design phase has been completed, it is recommended that the costs for the two phases be separated. If outside funding is necessary to complete a planned wetland project, consider separate grant applications for the two phases.

As shown in **Table 8** both phases have various associated tasks, all of which take time and involve costs. The costs associated with the various tasks will be influenced by:

- whether school staff or outside professionals are utilized to complete them;
- the size and complexity of the project;
- the earth-moving volumes and disposal costs for excavated materials;
- site specific conditions associated with the project location;
- distances of the project location from the various sources or disposal locations of materials; and
- meteorological conditions.

In spite of the impossibility of reliably estimating costs of the various tasks in **Table 8**, possible cost ranges are given for the tasks listed, given the limitations cited and assuming that all work is completed by nonunion professionals. It should be understood that actual costs may differ markedly because of the above listed effects.

INFILTRATION CONTROL WITH LINERS

On page 44, methods were cited to reduce water loss through infiltration. Soil compaction, which was used in the Construction Phase in **Table 8**, is the least expensive approach. Daily rental of a vibrator compactor and an infiltrometer may cost as little as $100. However, soil compaction may not be as effective as the use of a liner.

If there is an accessible source of clay, lining the planned wetland with at least six inches of clay would be the next least expensive approach for eliminating infiltration. In addition to the necessary volume of clay, a piece of equipment (i.e., a loader with a dozer bucket) would be needed to distribute and grade the clay throughout the planned wetland.

If the use of a PVC liner or a geosynthetic clay liner (e.g., Claymax 200R) is selected, check with the manufacturer to obtain an estimate for material (liner plus seam tape) and labor costs. As of 1998, labor and material cost estimates for the installation of these liners were:

$1.78/sq ft (20 mil PVC) and

$3.45/sq ft (Claymax).

Aquascape Designs, Inc. (1130 Carolina Drive, Unit C, West Chicago, IL 60185; 1-800-306-6227) sells only to wholesalers, but does installations. They use 45 mil EPDM (ethylene propylene diene monomer) as a liner and deal mostly with small residential ponds.

LANDSCAPING COSTS

Landscaping costs are determined by the designed quantity of plant materials (including fertilizer), the cost of planting materials, and the installation costs.

Quantity of Plant Materials

Landscaping costs may be kept at a minimum by ensuring that the specified plant quantities meet

Table 8 The Various Tasks and Possible Cost Ranges Associated with the Design and Construction Phases of Planned Wetland Projects

Tasks	Possible Cost Range
DESIGN PHASE	
Setting project goals	School staff
Setting hydroperiod, water depths, and conceptual design	School staff
Site selection:	
Collection and analysis of existing data for up to two site locations (tax maps, NWI maps, soil surveys, presence of biological benchmarks, etc.)	$350 to $1,500
Wetland delineation/assessment, if wetland is present	$250 to $750
Hydrological analysis and calculations for natural hydrology (includes stream gauging and piezometer monitoring during growing season)	$350 to $2,500
Soil analysis and infiltration testing at design elevations of planned wetland	$350 to $750
Topographic survey of planned wetland site (<1 acre in size)	$400 to $1,000
Grading and planting plans and specifications (<1 acre in size)	$1,500 to $3,000
Sediment and erosion control plans and specifications (<1 acre in size)	$800 to $1,000
CONSTRUCTION PHASE	
Earth-moving, excavation, and disposal of soils (<3 acre-feet of excatated soils, all stockpiled on site; infiltration control through soil compaction; no topsoil required)	$8,000 to $25,000
Constructing water control structures, and fish and wildlife features	$1,500 to $2,500
Planting and seeding (less than one-half acre)	$3,000 to $6,000
Maintenance:	
Watering (less than one-half acre of plants up to twice per growing season)	$0 to $3,000
Control of invasive vegetation and problem animals (yearly)	$0 to $1,500
Replanting	$0 to $250
Debris removal	$0 to $500
Structural and sediment erosion repairs	$0 to $1,000

Table 9 Recommended Planting Densities (On-center Spacing) to Achieve Uniform Ground Coverage by Herbaceous Plants in the Indicated Time

Rate of Spread	On-center Spacing Required for Coverage in:		
	1 Year	2 Years	3 Years
Rapid (>1 foot/year)	2 feet	4 feet	6 feet
Moderate (±0.5 foot/year)	1 foot	2 feet	3 feet
Slow (<0.2 foot/year)			0.5 foot

the needs of the project. **Table 9** provides the recommended planting densities for achieving uniform ground coverage of herbaceous plants with different rates of spread. **Table 10** and **Equation 4.4** (page 61) give the number of planting units, given planting densities (on-center distances) and planting areas. **Table 11** gives the unit quantity of fertilizer for the different types of plant materials, and **Equation 4.5** (page 61) gives the total pounds of fertilizer for a given number of planting units. (See pages 61-62.)

Be certain not to order overly large quantities of plant materials. For example, to achieve uniform ground cover of rapidly spreading herbaceous plants in one growing season, planting at a density of two feet on-center may be specified. However, should specifications require planting these species at a density of 0.5 foot on-center, rather than two feet on-center, the costs of planting materials and labor would increase unnecessarily by a factor of sixteen. Consequently, make certain not to specify planting more densely than needed.

Table 10 The Number of Planting Units as a Function of Planting Density (On-center Spacing)

In On-center Spacing is:	The Number of Planting Units will be:
0.5 foot (usually herbaceous)	(area in square feet) (4)
1 foot (usually herbaceous)	(area in square feet) (1)
1.5 feet (usually herbaceous)	(area in square feet) (0.444)
2 feet (usually herbaceous)	(area in square feet) (0.250)
3 feet (usually herbaceous)	(area in square feet) (0.110)
4 feet (usually herbaceous)	(area in square feet) (0.062)
5 feet (usually herbaceous)	(area in square feet) (0.040)
6 feet (usually shrubs)	(area in square feet) (0.028)
7 feet (usually shrubs)	(area in square feet) (0.020)
8 feet (usually shrubs)	(area in square feet) (0.016)
9 feet (usually trees)	(area in square feet) (0.012)
10 feet (usually trees)	(area in square feet) (0.010)
11 feet (usually trees)	(area in square feet) (0.008)

Table 11 Recommended Quantities of Fertilizer for Various Herbaceous (H) and Woody (W) Plant Materials

Type of Plant Material	Quantity of Osmocote	Quantity of Agriform
dormant propagule (H)	15 g (0.5 oz)	1 tablet
peat pot (H)	30 g (1 oz)	3 tablets
seedling plug (H & W)	15 g (0.5 oz)	1 tablet
1 quart container (H&W)	30 g (1 oz)	3 tablets
1 gallon container (W)	90 g (3 oz)	9 tablets
tree or shrub of given height	check manufacturers label	check manufacturers label

Suppliers should be contacted to obtain current prices for Osmocote® and Agriform® fertilizers. However, for cost estimates consider using these 1998 prices:

$1.20/lb for Osmocote®

$0.07/tablet for Agriform®.

Price adjustments should be made annually to the above values.

Costs of Plant Materials

The least expensive wetland plant materials, other than seed, are bare root (woody) and dormant propagules (herbaceous). These must be planted during the dormant season. These materials, while relatively inexpensive, have the poorest survival rates.

Plant materials that can be planted at any time of the year are containerized, either in peat pots or plastic pots. While these materials are most expensive, they offer the distinct advantage of providing the highest survival rates and no restrictions on planting times. Because of their advantages, containerized plant materials are recommended to be specified on all planned wetland projects. **Table 12** provides general cost ranges for containerized herbaceous and woody plant materials. These costs are fairly typical of those throughout the wetland nursery industry. However, special rates for certain materials are always available.

Seeding is not recommended as a method of establishing wetland plants. Although there are a few examples of successful seeding of wetland

Table 12 Price Ranges for Herbaceous (H) and Woody (W) Plant Materials that can be Planted During Any Time of the Year and that Have the Highest Survival Rates (1998 Prices)

Plant Material	Cost Range[1]
1.75 inch peat pot (H)	$0.09 to $0.65
1.5 inch x 6 inch catity plug (H & W)	$0.60 to $0.85
1 quart plastic pot (H)	$1.60 to $1.80
18 to 24 inches (W)	$3.00 to $7.00
2 to 3 feet (W)	$5.00 to $10.00
3 to 4 feet (W)	$7.00 to $12.00
4 to 5 feet (W)	$9.00 to $15.00

[1]Most nurseries offer quantity discounts. Contact local nurseries for exact pricing.

Table 13 Approximate Hours for the Installation of Wetland Plant Material

Items to be Planted	Number Planted per Worker Hour
1.5 to 1.75 inch peat pots, where dry	125
1.5 to 1.75 inch peat pots, under water	40
cavity plugs, where dry	140
cavity plugs, under water	45
1 quart pots, where dry	60
1 quart pots, under water	20
containerized trees 2 to 5 feet tall, with tractor and auger	15 to 50
containerized trees 2 to 5 feet tall, by hand	5 to 15

herbaceous plants, they are limited and difficult to replicate.

Before deciding to establish wetland plants by seeding:

- a reliable source of seeds must be located;
- the proper seed collection time, processing, and storage to ensure breaking of dormancy and maximum seed viability must be known;
- the proper seeding techniques and requirements must be known; and
- the proper hydrology requirements during seeding and after seeding for optimal seed germination and seedling emergence must be known.

If seeding is specified and unless it is known to the contrary, the time of seeding should be at the beginning of the growing season so as to provide the maximum time for the seedlings to mature before freezing temperatures or the dormant season commences.

Installation Costs

The actual planting of a wetland often involves difficulties not usually associated with landscape work. Therefore, the costs of installing wetland plant materials are often greater than those for other landscape projects. The usual formula for the labor cost of landscaping (total cost = three times the plant costs) will often prove insufficient for wetland planting.

Although planting in the dry is always preferred over planting in the wet, in many instances plants will have to be installed underwater. This will substantially increase the time required for planting.

In other cases, they will have to be installed by hand instead of with power augers because of site conditions. If it is necessary to pump water out of a site in order to properly install the plant materials followed by pumping the water back into the site, this should be figured into the planting costs.

Because of the varying labor costs, **Table 13** provides work hours generally needed to install wetland plant materials.

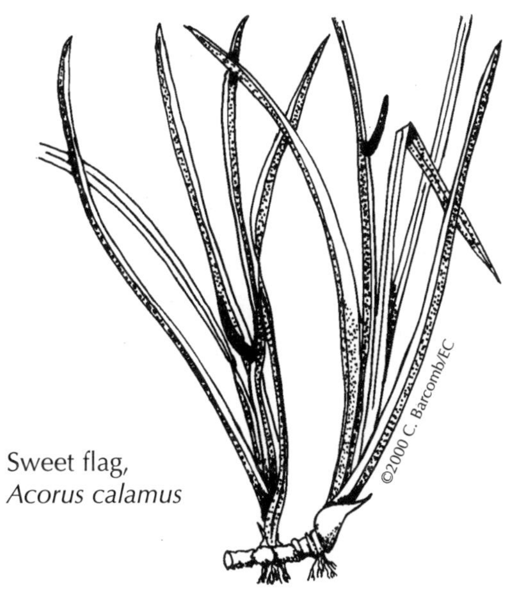

Sweet flag, *Acorus calamus*

7 COMMON WETLAND PLANTS

This chapter provides the following information on the dominant wetland plant species in the eastern United States (central and western species will be included in later editions):

- Characteristics
- Hydrology
- Habitat
- Notes
- Wildlife Benefits

The hydroperiods listed under the Hydrology information for each plant in nontidal wetlands is one of the four hydroperiods provided in the *Corps of Engineers Wetlands Delineation Manual* (1987). These hydroperiods refer to the duration of inundation and/or soil saturation that occurs during the growing season and are described on page 7. The wetland indicator status (pages 5-6) is indicated for each species. The addition of a plus after the indicator status indicates that the species falls on the wet side of the category. A minus indicates that the species is on the dry side of the category. Other hydrology information given for the plants is their maximum depth of flooding tolerance.

It may be helpful to remember that seasonally and irregularly flooded wetlands contain plant species that adapt to droughty conditions.

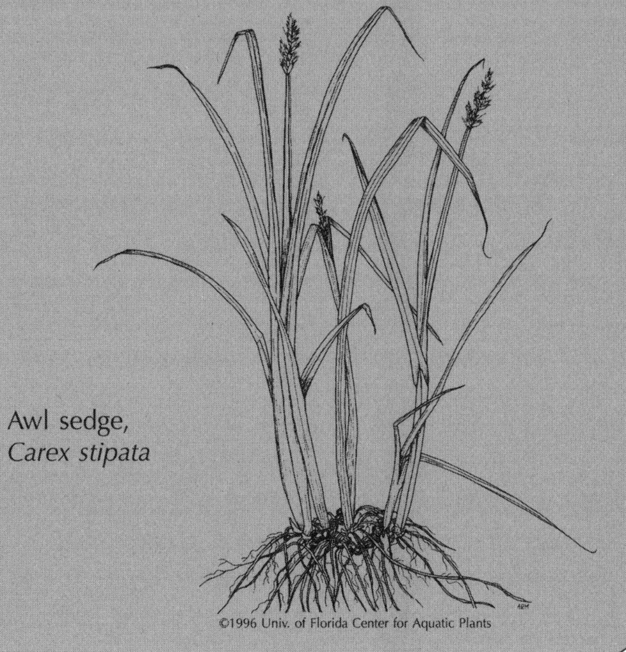

Awl sedge, *Carex stipata*

ACER RUBRUM

RED MAPLE

Characteristics

- deciduous tree, 75 to 100 feet in height

Hydrology

- indicator status: facultative to facultative wet plus (FAC to FACW+)
- salinity: no tolerance
- tidal zone: above spring tide elevation
- nontidal regime: irregularly to seasonally inundated or saturated, tolerates drought

Habitat

- fresh tidal marsh or swamp, nontidal marsh or swamp, alluvial woods, moist uplands

Notes

- tolerates partial shade; small transplants do not tolerate standing water; pH preference is 4.5 to 6.5

Wildlife Benefits

- nesting, cover: American robin, prairie warbler, American goldfinch
- seeds: water birds, bobwhite quail, cardinal, pine siskin, evening grosbeak, American goldfinch, yellow-bellied sapsucker
- twigs and foliage: hoofed browsers

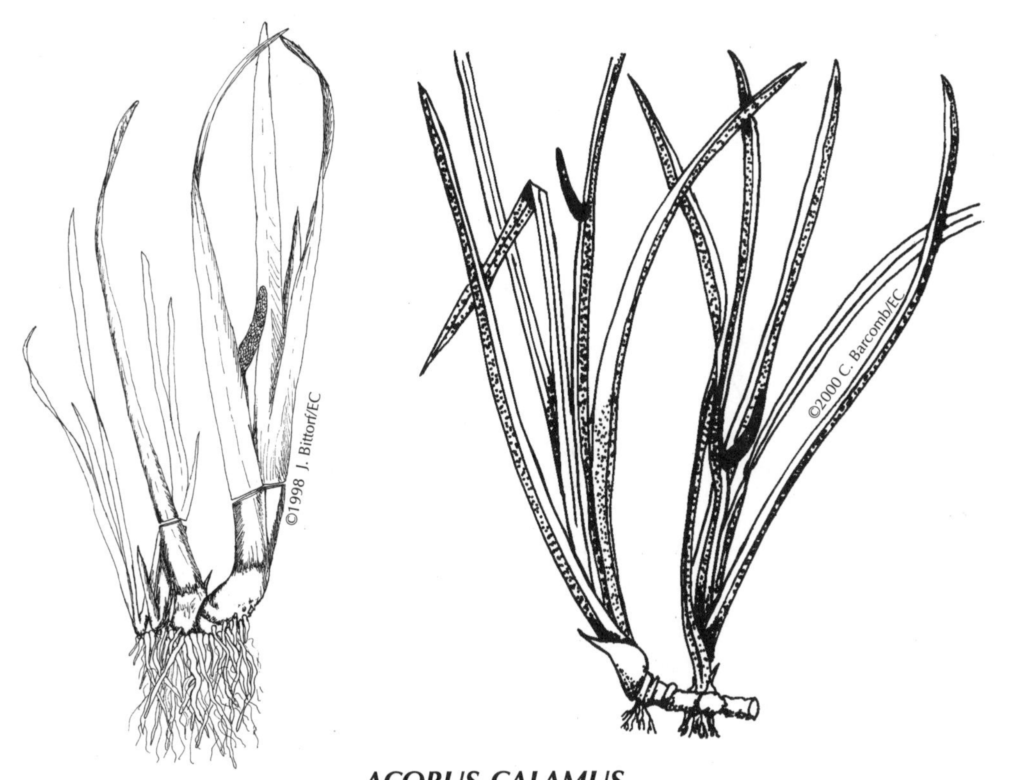

ACORUS CALAMUS

SWEET FLAG

Characteristics
- herbaceous, perennial, nonpersistent, 1 to 4 feet in height
- rate of spread: moderate; 0.5 ± 0.1 feet per year by rhizome (in unconsolidated sediments)

Hydrology
- indicator status: obligate (OBL)
- salinity: up to approximately 5 ppt
- tidal zone: from MHW to spring tide elevation
- nontidal regime: regularly to permanently inundated to a depth of 0.5 foot or saturated

Habitat
- fresh to brackish tidal marshes, nontidal marshes, wet meadows

Notes
- tolerates partial shade; soil stabilizer; tolerates low pH

Wildlife Benefits
- food (rhizomes) & cover: many species of waterfowl, muskrat

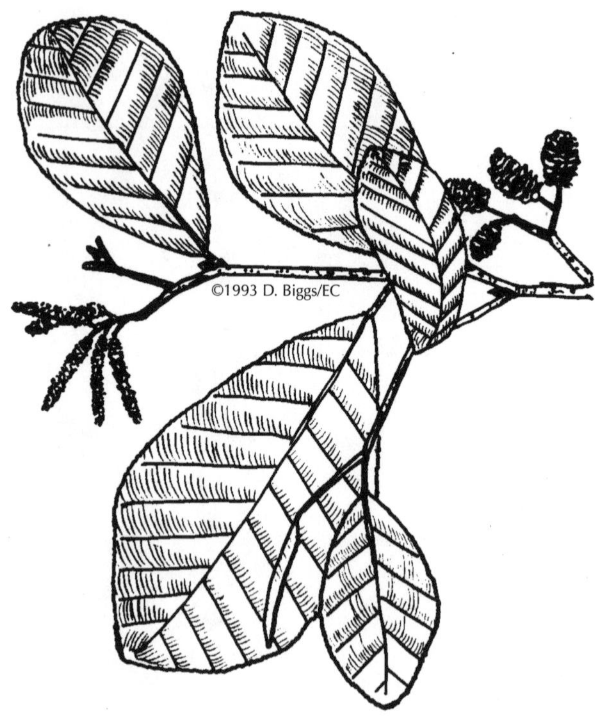

ALNUS SERRULATA

HAZEL, SMOOTH, AND TAG ALDER

Characteristics

- deciduous shrub, 12 to 20 feet in height

Hydrology

- indicator status: obligate (OBL)
- salinity: no tolerance
- tidal zone: above MHW to upland
- nontidal regime: seasonally to regularly inundated to a depth of 3 inches or saturated

Habitat

- fresh tidal marshes, nontidal marshes, swamps and forested wetlands (in open areas)

Notes

- requires full sun, fixes nitrogen, pH range is 5.5 to 7.5

Wildlife Benefits

- nesting, cover: American woodcock, willow flycatcher, fox sparrow, alder flycatcher, yellow warbler, Wilson's warbler, red-winged blackbird, rusty blackbird, ruffed grouse, beaver
- seeds: mallard, American wigeon, green-winged teal, bufflehead, turkey, common redpoll, pine siskin, American goldfinch, muskrat, beaver, cottontail rabbit, white-tailed deer, woodcock, ruffed grouse, many songbirds

ANDROPOGON VIRGINICUS

BROOM SEDGE

Characteristics

- herbaceous/perennial/persistent, 1 to 3 feet in height
- rate of spread: slow, <0.2 foot per year by rhizomes (in unconsolidated sediments); a nonaggressive bunchgrass

Hydrology

- indicator status: not listed; however, has been grown in nursery in permanently saturated conditions for several growing seasons.
- salinity: freshwater only, will tolerate saltwater spray
- nontidal regime: irregularly to seasonally inundated or saturated; tolerates drought

Habitat

- wet meadows, forested wetland (in open areas and along edges), transition to uplands, buffer zone

Notes

- requires full sun; suitable as a nurse crop for forested wetlands

Wildlife Benefits

- seeds: finch, junco, field sparrow, and tree sparrow
- plants: antelope, bison, white-tailed deer

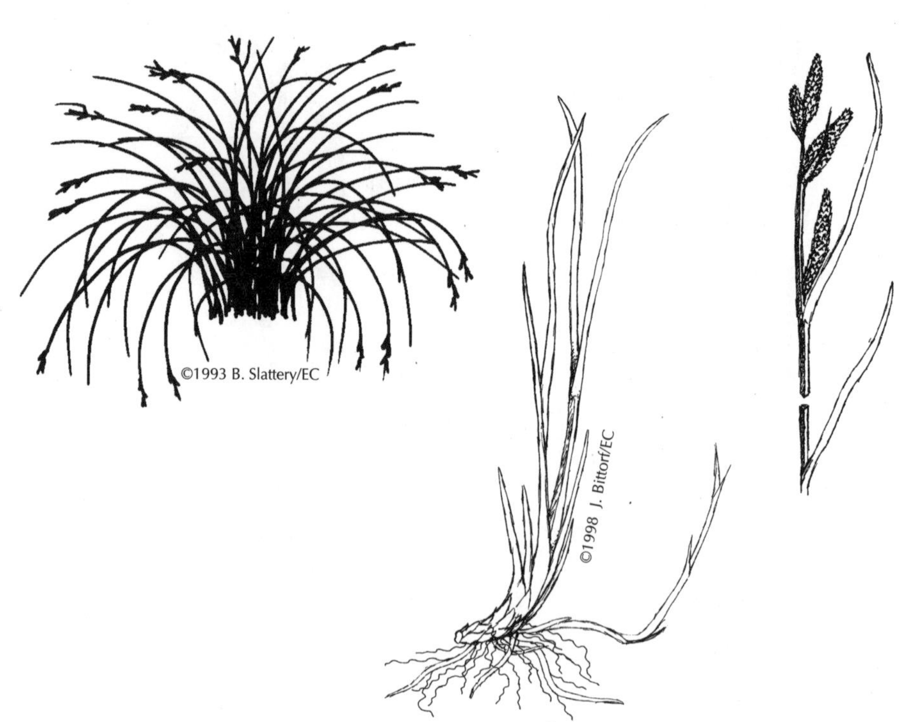

CAREX STRICTA

TUSSOCK AND UPTIGHT SEDGE

Characteristics

- herbaceous, perennial, semi-persistent, up to 3.5 feet in height
- rate of spread: moderate, 0.5 ± 0.1 foot per year by rhizomes (in unconsolidated sediments)

Hydrology

- indicator status: obligate (OBL)
- salinity: no tolerance
- nontidal regime: seasonally regularly inundated, or permanently inundated up to a depth of 0.5 foot or saturated

Habitat

- nontidal marshes, shrub swamps and forested wetlands (in open areas), wet swales

Notes

- requires full sun; tolerates low pH; forms raised clumps or "tussocks"

Wildlife Benefits

- cover: many small animals
- food (rhizomes, leaves, shoots): beaver, muskrat, deer, moose, elk
- seeds: sora rail, yellow rail, grouse, snipe, seed-eating songbirds, swamp sparrow, tree sparrow, Lincoln sparrow, snow bunting, larkspur, redpoll, ruffed grouse, black duck

CEPHALANTHUS OCCIDENTALIS

BUTTONBUSH

Characteristics
- deciduous shrub, 6 to 12 feet in height

Hydrology
- indicator status: obligate (OBL)
- salinity: resistant; tolerates infrequent brackish-water flooding
- tidal zone: near MHW to spring tide elevation
- nontidal regime: irregularly to permanently inundated to a depth of 3 feet or saturated

Habitat
- fresh tidal marshes, nontidal marshes, scrub-shrub wetlands, forested wetlands, borders of streams, lakes, and ponds

Notes
- tolerates full shade to full sun; flowers best in full sun; will adapt to upland conditions; roots readily from cuttings; fairly insensitive to disease, and to insect, wind, or ice damage

Wildlife Benefits
- seeds: mallard, wigeon, shoveler, wood duck, teals
- nesting: Virginia rail, red-winged blackbird, song sparrow
- nectar: ruby-throated hummingbird
- attracts: muskrat, beaver

CHAMAECYPARIS THYOIDES

ATLANTIC WHITE CEDAR, FALSE CYPRESS, SOUTHERN WHITE CEDAR

Characteristics

- evergreen coniferous tree, 40 to 50 feet in height

Hydrology

- indicator status: obligate (OBL)
- salinity: no tolerance
- nontidal regime: irregularly to permanently inundated up to a depth of 0.5 foot or saturated

Habitat

- forested wetlands, shrub bogs, stream edges

Notes

- prefers full sun; cannot compete with hardwood species and is readily outgrown; usually found growing on hummocks; pH preference is 3.0 to 5.0; can adapt to upland conditions

Wildlife Benefits

- winter browse and food: white-tailed deer, pine siskin
- seedlings: cottontail rabbit, meadow mouse
- community: bear, beaver, otter, deer, parula warbler, prairie warbler, prothonotary warbler, hooded warbler, worm-eating warblers, cooper's hawk, red-shouldered hawk, barred owl, oven bird, yellowthroat, etc.

CORNUS AMOMUM

SILKY DOGWOOD

Characteristics

- deciduous shrub, 6 to 12 feet in height

Hydrology

- indicator status: facultative wet (FACW)
- salinity: no tolerance
- nontidal regime: irregularly to seasonally inundated at surface (< 0.5 foot depth) or saturated

Habitat

- forested wetlands (in open areas), shrub wetlands, stream banks, moist woods

Notes

- requires full sun; will adapt to upland conditions; fairly insensitive to insect, wind, and ice damage; pH preference is 6.1 to 7.5 (tolerates pH of 8.5)

Wildlife Benefits

- fruit: waterfowl, downy woodpecker, cedar waxwing, common flicker, eastern bluebird
- cover and nesting: gray catbird
- food: white-tailed deer, wild turkey, beaver, pileated woodpecker, ruffed grouse, bobwhite quail, ring-necked pheasant, cottontail rabbit, woodchuck, raccoon, squirrel

CORNUS SERICEA

RED-OSIER DOGWOOD

Characteristics

- deciduous shrub, 8 to 15 feet in height
- rate of spread: < 0.2 foot per year by stolons

Hydrology

- indicator status: facultative to facultative minus (FAC to FAC–)
- salinity: no tolerance
- tidal zone: throughout spring tide elevations of tidal freshwater wetlands
- nontidal regime: irregularly to seasonally inundated to < 0.5 foot depth or saturated

Habitat

- fresh tidal marsh or swamp, nontidal swamp, shrub wetlands, forested wetlands, stream banks

Notes

- tolerates partial shade; will adapt to upland conditions; formerly *Cornus stolonifera*

Wildlife Benefits

- cover and nesting: American goldfinch
- food: eastern kingbird, brown thrasher, purple thrush, ring-necked pheasant, white-tailed deer, wild turkey, beaver, ruffed grouse, bobwhite quail, cottontail rabbit, woodchuck, raccoon

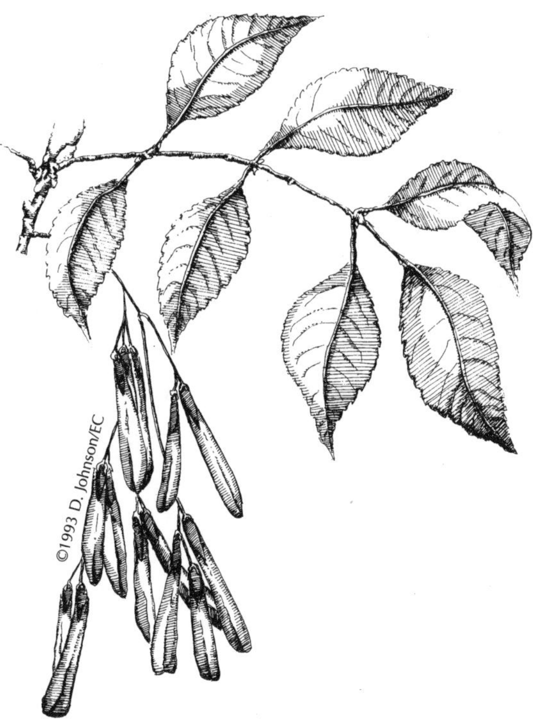

FRAXINUS PENNSYLVANICA

GREEN ASH

Characteristics

- deciduous tree, 50 to 75 feet in height

Hydrology

- indicator status: facultative wet (FACW)
- salinity: resistant; tolerates infrequent brackish water flooding
- tidal zone: above MHW to upland elevations
- nontidal regime: regularly, seasonally, and irregularly inundated to < 0.5 foot depth or saturated

Habitat

- tidal freshwater forested wetlands, nontidal forested wetlands, swamps

Notes

- tolerates partial shade; will adapt to upland conditions; susceptible to wind or ice damage, fairly insensitive to disease and insect damage; tolerates drought

Wildlife Benefits

- seeds: wood duck, bobwhite quail, red-winged blackbird, cardinal, purple finch, pine grosbeak
- sap: yellow-bellied sapsucker
- cover and nesting: mourning dove, evening grosbeak
- browse: white-tailed deer

HIBISCUS MOSCHEUTOS

MARSH HIBISCUS, ROSE MALLOW

Characteristics
- woody, perennial, persistent
- 4 to 7 feet in height, shrub-like with new growth coming from below ground each year
- rate of spread: slow, < 0.2 foot per year (in unconsolidated sediments)

Hydrology
- indicator status: obligate (OBL)
- salinity: tolerates up to approximately 15 ppt
- tidal zone: upper 20% of intertidal zone
- nontidal regime: irregularly to regularly inundated to a depth of 3 inches or saturated

Habitat
- tidal fresh, brackish, and salt marshes, nontidal marshes (occasionally)

Notes
- requires full sun, will adapt to upland conditions
- nectar: ruby-throated hummingbird

Wildlife Benefits
- food, cover, nesting: limited

ILEX VERTICILLATA

COMMON WINTERBERRY, WINTERBERRY HOLLY

Characteristics
- deciduous shrub, 6 to 12 feet in height

Hydrology
- indicator status: facultative wet plus (FACW+)
- salinity: no tolerance
- tidal zone: above MHW to upland elevations
- nontidal regime: irregularly to seasonally inundated up to depth of 0.5 foot or saturated

Habitat
- fresh tidal swamps, shrub swamps, forested wetlands

Notes
- tolerates full to partial shade; male and female required for berries (other *Ilex* spp. will pollinate)
- fairly insensitive to disease or to insect, wind, or ice damage
- pH preference is 4.5 to 5.0 (tolerates up to 8.0); will adapt to upland conditions

Wildlife Benefits
- food and berries: mockingbird, catbird, brown thrasher, hermit thrush, cottontail rabbit, raccoon, white-footed mouse, squirrel, ruffed grouse, ring-necked pheasant
- browse: white-tailed deer

IRIS VERSICOLOR

BLUE FLAG, POISON FLAG, CLAJEU

Characteristics

- herbaceous, perennial, nonpersistent up to 4 feet in height
- rate of spread: slow, <0.2 foot per year (in unconsolidated sediments)

Hydrology

- indicator status: obligate (OBL)
- salinity: no tolerance
- tidal zone: MHW to spring tide elevations
- nontidal regime: regularly to permanently inundated to a depth of 0.5 foot or saturated

Habitat

- fresh tidal marshes, nontidal marshes, wet meadows and shores, swamps and forested wetlands

Notes

- requires full sun for flowers, but will tolerate partial shade; remains in clumps

Wildlife Benefits

- food: muskrat, wildfowl (probably not seeds), marsh birds
- cover: persists under heavy grazing

LEERSIA ORYZOIDES

RICE CUTGRASS

Characteristics

- herbaceous, perennial, nonpersistent, up to 5 feet in height
- rate of spread: moderate, 0.5 ± 0.1 foot per year (in unconsolidated sediments)

Hydrology

- indicator status: obligate (OBL)
- salinity: no tolerance
- tidal zone: upper 10% of intertidal zone
- nontidal regime: irregularly, seasonally, regularly, or permanently inundated to a depth of 0.5 foot or saturated

Habitat

- fresh tidal marshes, nontidal marshes, wet meadows, ditches and shores, swamps and forested wetlands

Notes

- tolerates full shade; effective for sediment stabilization and erosion control

Wildlife Benefits

- seeds: many species of ducks, swamp sparrow, tree sparrow, sora rail
- food source: rails, waterfowl, herons, other birds (as habitat for invertebrates); and for raptors, herons, mammals, etc. (as habitat for fish, reptiles, amphibians)

LINDERA BENZOIN

COMMON SPICEBUSH

Characteristics

- deciduous shrub, 6 to 12 feet in height

Hydrology

- indicator status: facultative wet minus (FACW–)
- salinity: resistant; tolerates infrequent brackish-water flooding
- tidal zone: not normally found
- nontidal regime: seasonally inundated to a depth of 0.5 foot or saturated

Habitat

- forested wetlands

Notes

- formerly *Benzoin aestivale*; tolerates full shade (while producing fruits); fairly insensitive to disease and insects, and to wind and ice damage, pH tolerance is 4.5 to 6.5

Wildlife Benefits

- berries: wood thrush, veery, ruffed grouse, bobwhite, ring-necked pheasant, common flicker, eastern kingbird, great crested flycatcher, gray catbird, American robin, hermit thrush, gray-cheeked thrush, red-eyed vireo, cardinal, white-throated sparrow, wild turkey
- browse: white-tailed deer

MYRICA PENSYLVANICA

BAYBERRY

Characteristics
- deciduous shrub, 6 to 12 feet in height

Hydrology
- indicator status: facultative (FAC)
- salinity: up to approximately 20 ppt
- tidal zone: above MHW to upland
- nontidal regime: irregularly to seasonally inundated at surface or saturated

Habitat
- tidal fresh, brackish, and salt marshes, tidal fresh and brackish swamps, nontidal marshes and swamps (infrequent)

Notes
- tolerates partial to full shade; nitrogen fixing; male and female required for seed production; fairly insensitive to disease, and to insect, wind, and ice damage; pH preference is 5.0 to 6.5

Wildlife Benefits
- berries: eastern meadowlark, white-eyed vireo, yellow-rumped warbler, tree swallow, red-winged blackbird
- winter food (berries): many songbirds, waterfowl, shorebirds, and marsh birds
- cover: many species

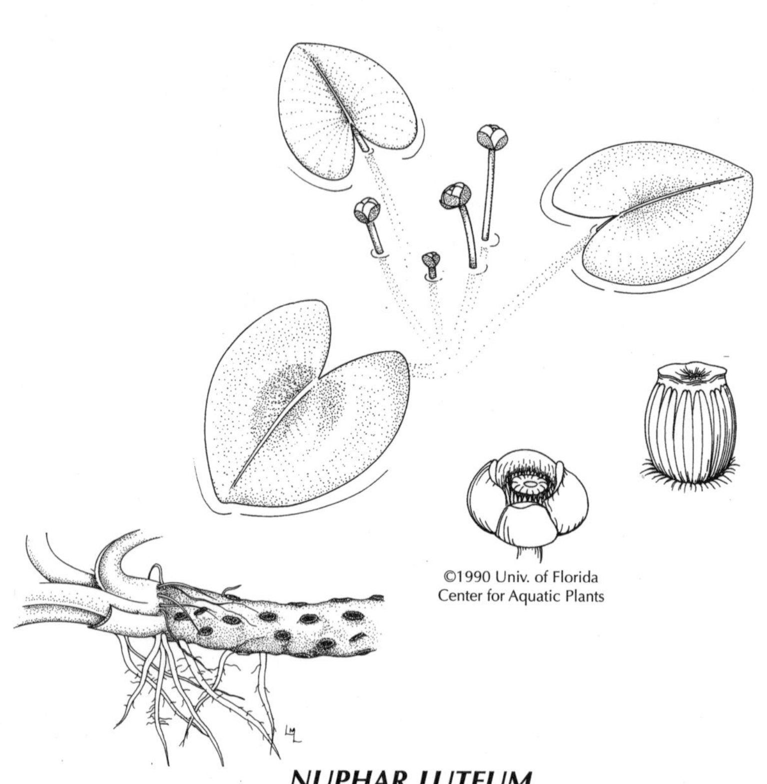

NUPHAR LUTEUM

SPATTERDOCK, YELLOW WATER LILY, COWLILY

Characteristics

- herbaceous, perennial, nonpersistent, floating, up to 16 inches in height
- rate of spread: slow, less than 0.2 foot per year by rhizome (in unconsolidated sediments)

Hydrology

- indicator status: obligate (OBL)
- salinity: resistant; tolerates infrequent brackish-water flooding
- tidal zone: about MLW to 3 feet below MLW
- nontidal regime: regularly to permanently inundated to depths of 3 feet

Habitat

- fresh tidal marshes, nontidal marshes, swamps, lakes and ponds

Notes

- requires full sun

Wildlife Benefits

- seed: ring-necked duck, wood duck, Florida duck
- food (plants): beaver, porcupine, deer, muskrat
- shade, shelter, and habitat for insects which are consumed by fish

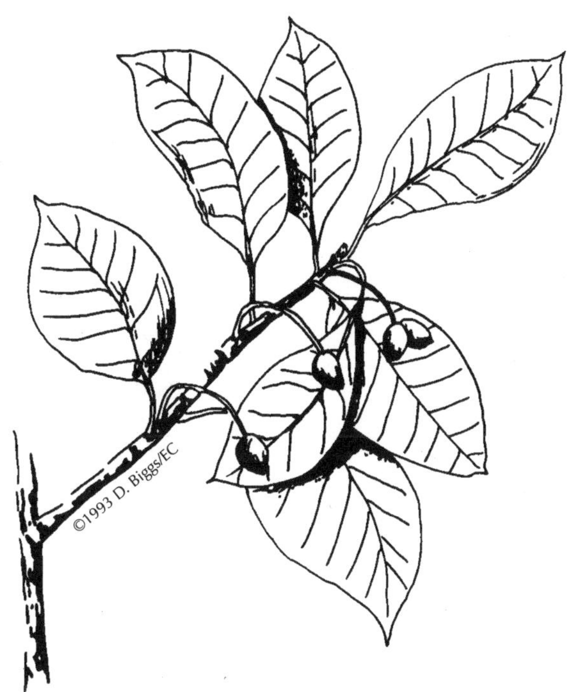

NYSSA SYLVATICA

BLACK GUM, BLACK TUPELO, SOUR GUM

Characteristics

- deciduous tree, 50 to 75 feet in height

Hydrology

- indicator status: facultative (FAC)
- salinity: resistant; tolerates infrequent brackish-water flooding
- tidal zone: above MHW to upland elevations in freshwater tidal marshes
- nontidal regime: irregularly to seasonally inundated to a depth of 0.5 foot or saturated

Habitat

- tidal freshwater swamps, forested wetlands

Notes

- tolerates drought; will adapt to upland conditions; pH preference is 6.1 to 6.5, male and female flowers on different trees; fairly insensitive to disease, and to insect, wind, and ice damage; provides distinctive fall foliage color

Wildlife Benefits

- food (fruit): wood duck, common flicker, cedar waxwing, summer tanager, redheaded woodpecker, aquatic furbearers
- hollow trunks: raccoon, owl
- browse: white-tailed deer

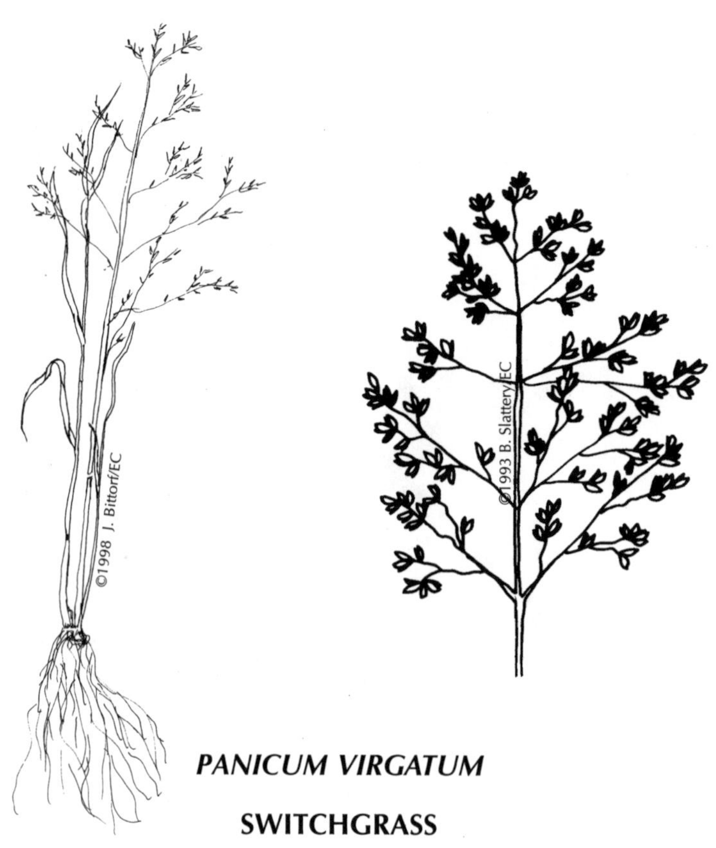

PANICUM VIRGATUM

SWITCHGRASS

Characteristics

- herbaceous, perennial, persistent up to 6 feet in height
- rate of spread: slow, < 0.2 foot per year by rhizomes (in unconsolidated sediments)

Hydrology

- indicator status: facultative (FAC)
- salinity: to approximately 10 ppt
- tidal zone: above MHW to upland in fresh or brackish water, and in salt marshes
- nontidal regime: irregularly to seasonally inundated to depth of 0.5 foot or saturated

Habitat

- tidal fresh, brackish, and salt marshes, nontidal marshes, wet meadows

Notes

- tolerates drought; will adapt to upland conditions; a non-turf forming bunch grass; recommended as a suitable transitional and buffer species; can be established by seeding; takes two growing seasons to flower

Wildlife Benefits:

- seeds and plants: Florida duck, teals, white-fronted goose, snow goose, baldpate duck, muskrat, rabbit, deer
- seeds: snipes, ground dove, quail, wild turkey, red-winged blackbird, cowbird, blue grosbeak, longspur, pyrrhuloxia, sparrows (tree, savannah, Lincoln, etc.), white-footed mouse

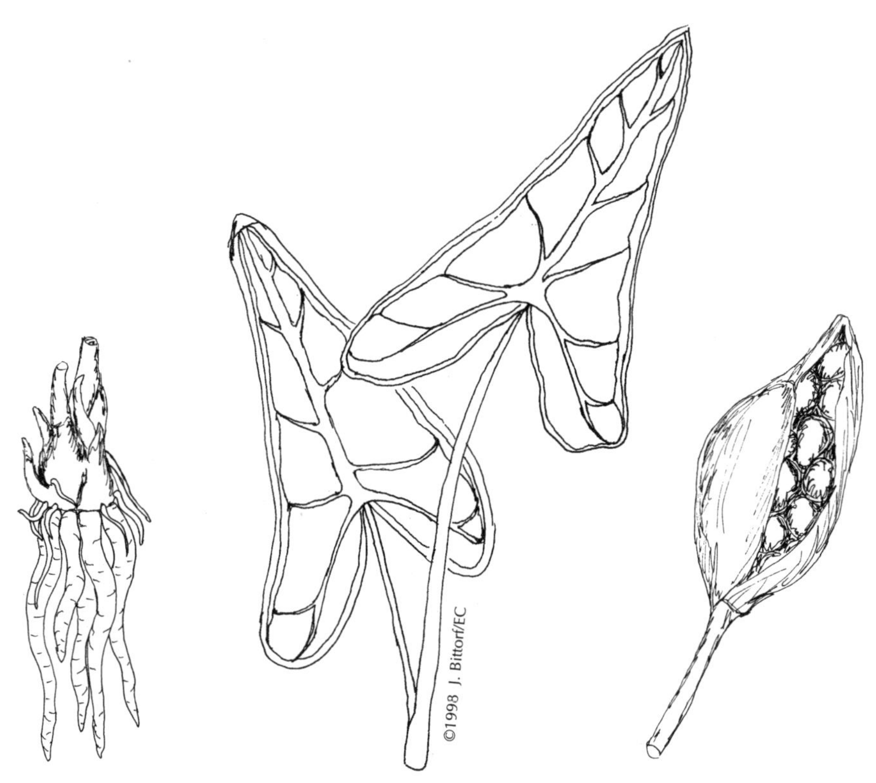

PELTANDRA VIRGINICA

ARROW ARUM, TUCKAHOE, WAMPEE, DUCK CORN

Characteristics

- herbaceous, perennial, nonpersistent, up to 2 feet in height
- rate of spread: slow, < 0.2 foot per year by bulb (in unconsolidated sediments)

Hydrology

- indicator status: obligate (OBL)
- salinity: no tolerance
- tidal zone: upper 50% of the intertidal zone
- nontidal regime: regularly to permanently inundated to a depth of 2 feet or saturated

Habitat

- fresh tidal marshes, nontidal marshes, swamps, ponds and lakes

Notes

- tolerates partial shade; foliage and rootstock NOT eaten by geese or muskrats

Wildlife Benefits

- seeds: wood duck, rail

POLYGONUM PUNCTATUM

MARSH OR DOTTED SMARTWEED, RED TOP

Characteristics

- herbaceous, perennial, nonpersistent, up to 4 feet in height

Hydrology

- indicator status: obligate (OBL)
- salinity: no tolerance
- tidal zone: upper 50% of intertidal zone
- nontidal regime: regularly to permanently inundated to a depth of 0.5 foot or saturated

Habitat

- fresh tidal marshes, nontidal marshes

Notes

- tolerates mildly alkaline soils

Wildlife Benefits:

- seeds: waterfowl, marsh birds, shorebirds, upland game birds, songbirds, aquatic furbearers, small mammals

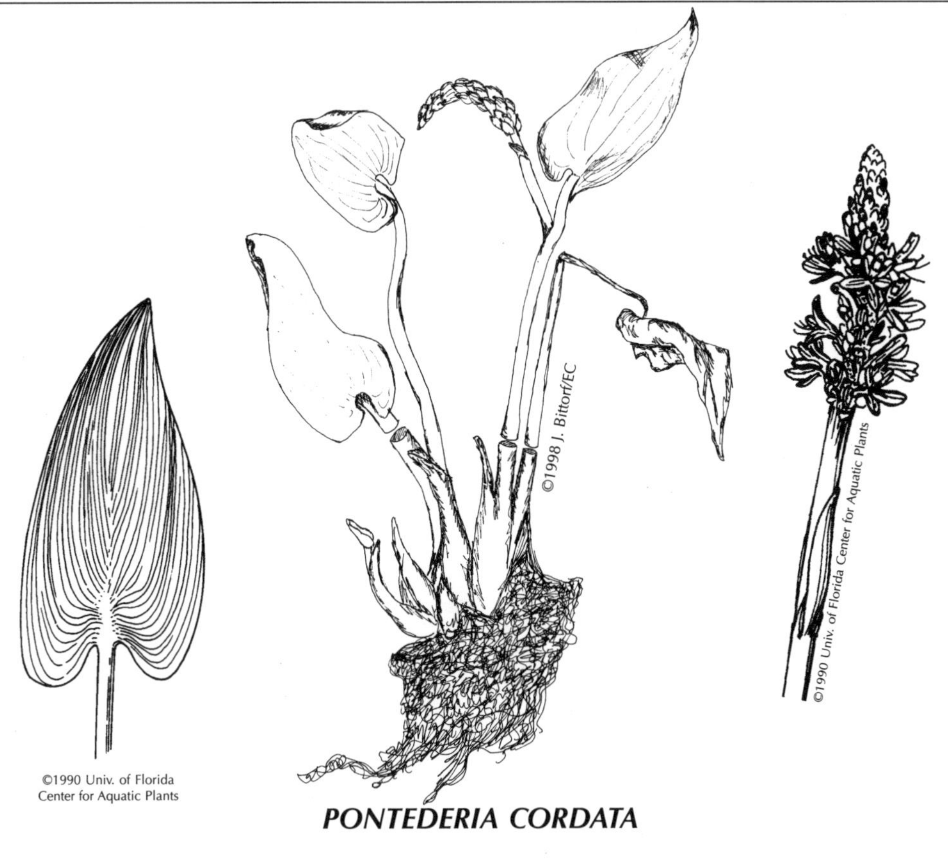

PONTEDERIA CORDATA

PICKERELWEED

Characteristics

- herbaceous, perennial, nonpersistent, up to 4 feet in height
- rate of spread: moderate, up to 0.5 foot per year by rhizome (in unconsolidated sediments)

Hydrology

- Indicator status: obligate (OBL)
- salinity: fresh to slightly brackish (1 to 2 ppt)
- tidal zone: upper 50% of the intertidal zone

Habitat

- fresh to slightly brackish marshes, nontidal marshes, ponds and lakes

Notes

- tolerates partial shade; susceptible to freeze kill

Wildlife Benefits

- seeds: mottled duck, other waterfowl
- foliage and root stock: Canada geese, muskrat

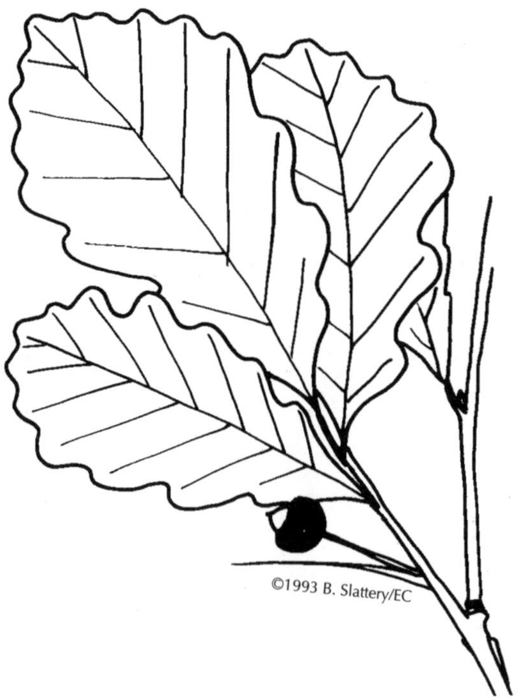

QUERCUS BICOLOR

SWAMP WHITE OAK

Characteristics

- deciduous tree, 75 to 100 feet in height

Hydrology

- indicator status: facultative wet plus (FACW+)
- salinity: resistant; tolerates infrequent flooding by brackish water
- nontidal regime: irregularly to seasonally inundated to a depth of 0.5 foot or saturated

Habitat

- forested wetlands

Notes

- tolerates drought; will adapt to upland conditions; tolerates partial shade; pH preference is 6.0 to 6.5; fairly insensitive to disease, and to insect, wind, and ice damage; susceptible to iron chlorosis

Wildlife Benefits

- acorns and buds: waterfowl (especially wood duck), marsh birds, shorebirds, upland game birds, other birds (particularly grackle, blue jay, brown thrasher, red-bellied woodpecker, redheaded woodpecker), raccoon, tree squirrel, eastern chipmunk, white-footed mouse

QUERCUS PALUSTRIS

PIN OAK, SPANISH OAK

Characteristics

- deciduous tree, 50 to 75 feet in height

Hydrology

- indicator status: facultative wet (FACW)
- salinity: resistant; tolerates infrequent flooding by brackish water
- nontidal regime: irregularly to seasonally inundated to a depth of 0.5 foot or saturated

Habitat

- forested wetlands, alluvial woods

Notes

- tolerates drought; will adapt to upland conditions; requires full sun; pH preference is 5.5 to 6.5; fairly insensitive to disease and to insect, wind, and ice damage; susceptible to iron chlorosis

Wildlife Benefits

- acorns and buds: wood duck, mallard, quail, wild turkey, ruffed grouse, bobwhite, blue jay, brown thrasher, rufous-sided towhee, deer, fox, opossum, raccoon
- cover and nesting: scarlet tanager, rose-breasted grosbeak

QUERCUS PHELLOS

WILLOW OAK

Characteristics

- deciduous tree, 40 to 60 feet in height

Hydrology

- indicator status: facultative plus (FAC+)
- salinity: no tolerance
- nontidal regime: irregularly to seasonally inundated to a depth of 0.5 foot or saturated

Habitat

- forested wetlands, floodplain and alluvial woods

Notes

- tolerates drought; will adapt to upland conditions; tolerates partial shade

Wildlife Benefits

- food, cover, nesting: wood duck, mallard, quail, wild turkey, deer, fox, opossum, raccoon, common grackle, ruffed grouse, green-winged teal, red-bellied woodpecker

SAGITTARIA LATIFOLIA

ARROWHEAD, DUCK POTATO, WAPATO

Characteristics

- herbaceous, perennial, nonpersistent, up to 4 feet in height
- rate of spread: rapid, >1 foot per year by underground runners (in unconsolidated sediments)

Hydrology

- indicator status: obligate (OBL)
- salinity: no tolerance
- tidal zone: middle 50% of intertidal zone
- nontidal regime: regularly to permanently inundated up to a depth of 2 feet or saturated

Habitat

- fresh tidal marshes, nontidal marshes, swamps, ponds and lakes

Notes

- tolerates partial shade; male and female flowers on separate plants; tolerates anoxic soils

Wildlife Benefits

- seeds, foliage, and tubers: geese, trumpeter swan, tundra swan, canvasback duck, gadwall duck, black duck, mallard, pintail duck, ring-necked duck, scaup, rails, muskrat, beaver

SALIX NIGRA

BLACK WILLOW

Characteristics

- deciduous tree, 35 to 50 feet in height
- rate of spread: slow, < 0.2 foot per year by suckers

Hydrology

- indicator status: facultative wet plus (FACW+)
- salinity: no tolerance
- nontidal regime: irregularly to regularly inundated to a depth of 0.5 foot or saturated

Habitat

- fresh tidal marshes and swamps, nontidal forested wetlands

Notes

- requires full sun; roots readily from cuttings; rapid grower; streambank stabilizer; susceptible to wind and ice damage; pH preference is 6.5 to 8.0

Wildlife Benefits

- fruit, buds, flowers: deer, beaver, squirrel, other small game, mallard, northern shoveler, wood duck, green-winged teal, turkey, ruffed grouse, bobwhite, ring-necked pheasant, mourning dove, many woodpeckers, tufted titmouse, Carolina wren, brown thrasher, hermit thrush, eastern meadowlark, rose-breasted grosbeak, pine grosbeak, common grackle
- cover and nesting: common crow, Baltimore oriole, scarlet tanager

SAMBUCUS CANADENSIS

ELDERBERRY, AMERICAN ELDER

Characteristics

- deciduous shrub, 6 to 12 feet in height
- rate of spread: slow, < 0.2 foot per year by suckers

Hydrology

- indicator status: facultative wet minus (FACW–)
- salinity: no tolerance
- tidal zone: above spring tide elevation
- nontidal regime: seasonally to irregularly inundated to a depth of 0.5 foot or saturated

Habitat

- fresh tidal marshes, nontidal marshes, forested wetlands, swamps

Notes

- tolerates full shade; tolerates drought; will adapt to upland conditions; pH preference is 6.1 to 7.5; susceptible to wind and ice damage (weak-wooded)

Wildlife Benefits

- fruit: mourning dove, yellow-bellied sapsucker, eastern kingbird, great crested flycatcher, eastern bluebird, American robin, starling, veery, blue jay, many woodpeckers and sparrows
- cover and nesting: alder flycatcher, mockingbird, gray catbird
- twigs, leaves: hoofed browsers

SCIRPUS CYPERINUS

WOOL GRASS

Characteristics

- herbaceous, perennial, persistent, 4 to 6.5 feet in height
- rate of spread: moderate, 0.5 foot per year by rhizome (in unconsolidated sediments)

Hydrology

- indicator status: facultative wet plus (FACW+)
- salinity: no tolerance
- nontidal regime: irregularly to seasonally inundated

Habitat

- fresh tidal marshes, nontidal marshes, swamps

Notes

- requires full sun

Wildlife Benefits

- cover and food (seeds or rhizomes): baldpate, black, canvasback, gadwall, mallard, mottled, pintail, redhead, ring-necked, ruddy, greater scaup, lesser scaup, shoveler, blue-winged teal, cinnamon teal, and green-winged teal ducks; Canada, tule, snow, and blue geese; trumpeter swan, sandhill crane, long-billed dowitcher, Hudsonian godwit, sora and Virginia rails, semipalmated sandpiper, snipe, quail, muskrat, fish
- nesting: bluegills, largemouth bass, and black crappie

SCIRPUS PUNGENS (S. AMERICANUS)

COMMON THREE SQUARE

Characteristics

- herbaceous, perennial, semi-persistent, up to 3 feet in height
- rate of spread: rapid, > 1 foot per year by rhizome (in unconsolidated sediments)

Hydrology

- indicator status: facultative wet plus (FACW+)
- salinity: up to 10 ppt
- tidal zone: upper 50% of intertidal zone
- nontidal regime: seasonally to permanently inundated to depths of 0.5 foot or saturated

Habitat

- fresh to brackish tidal marshes, nontidal marshes,
- shores of rivers, lakes, and ponds

Notes

- tolerates drought; requires full sun; soil stabilizer through root mat development

Wildlife Benefits

- same as *S. cyperinus*

SCIRPUS VALIDUS

SOFT STEM BULRUSH

Characteristics

- herbaceous, perennial, nonpersistent, up to 10 feet in height
- rate of spread: rapid, > 1 foot per year (in unconsolidated sediments)

Hydrology

- indicator status: obligate (OBL)
- salinity: up to about 5 ppt
- tidal zone: middle 50% of intertidal zone
- nontidal regime: regularly to permanently inundated to a depth of 1 foot or saturated

Habitat

- slightly brackish and fresh tidal marshes, nontidal marshes, shores

Notes

- requires full sun; soil stabilizer through root mat development

Wildlife Benefits

- same as for *S. cyperinus*

SPARGANIUM AMERICANUM

EASTERN OR LESSER BUR-REED

Characteristics

- herbaceous, perennial, nonpersistent, up to 3.5 feet in height
- rate of spread: rapid, > 1 foot per year (in unconsolidated sediments)

Hydrology

- indicator status: obligate (OBL)
- salinity: no tolerance
- nontidal regime: regularly to permanently inundated to depth of 0.5 foot or saturated

Habitat

- nontidal marshes, shores

Notes

- tolerates partial shade

Wildlife Benefits

- seeds: mallard, wood, canvasback, ring-necked, scaup, and teal ducks; tundra swans, muskrats, beavers
- food (plants): muskrat, Canada goose

SPARTINA ALTERNIFLORA

SMOOTH CORDGRASS, SALT MARSH CORDGRASS

Characteristics

- herbaceous, perennial, semi-persistent, 3 to 7 feet in height
- rate of spread: rapid, > 1 foot per year by rhizome (in unconsolidated sediments)

Hydrology

- indicator status: obligate (OBL)
- salinity: up to 35 ppt
- tidal zone: upper 50% of the intertidal zone
- nontidal regime: none, found only in tidal wetlands

Habitat

- tidal brackish and salt marshes

Notes

- requires full sun; soil stabilizer through root mat development; extremely sensitive to anoxic soils and will not survive in poorly drained areas; plants native to Maine and Massachusetts are known to be genetically isolated from plants native to the range from Connecticut to the Carolinas

Wildlife Benefits

- seeds, foliage, rhizomes: black ducks, Canada geese, snow geese, rails, seaside and sharp-tailed sparrows, muskrats

SPARTINA PATENS

SALT MARSH HAY, SALTMEADOW CORDGRASS, HIGHWATER GRASS

Characteristics

- herbaceous, perennial, persistent, 1 to 3 feet in height
- rate of spread: moderate, 0.5 foot per year by rhizome (in unconsolidated sediments)

Hydrology

- indicator status: facultative wet plus (FACW+)
- salinity: up to 35 ppt
- tidal zone: above MHW to upland
- nontidal regime: none, found only in tidal wetlands

Habitat

- tidal brackish and salt marshes, coastal dunes

Wildlife Benefits

- seeds, foliage, rhizomes: Canada and snow geese, black duck, sparrows, rails

THUJA OCCIDENTALIS

NORTHERN WHITE CEDAR, ARBOR VITAE

Characteristics

- evergreen tree, 50 to 75 feet in height

Hydrology

- indicator status: facultative wet (FACW)
- salinity: no tolerance
- tidal zone: not found in tidal wetlands
- nontidal regime: seasonally or regularly inundated to a depth of 0.5 foot or saturated

Habitat

- seasonally flooded forested wetlands, swamps, bogs

Notes

- tolerates partial shade; susceptible to wind and ice damage (weak-wooded); pH preference is 6.0 to 8.0

Wildlife Benefits

- buds: red squirrel
- seeds: pine siskin
- browse: snowshoe hare, white-tailed deer, moose

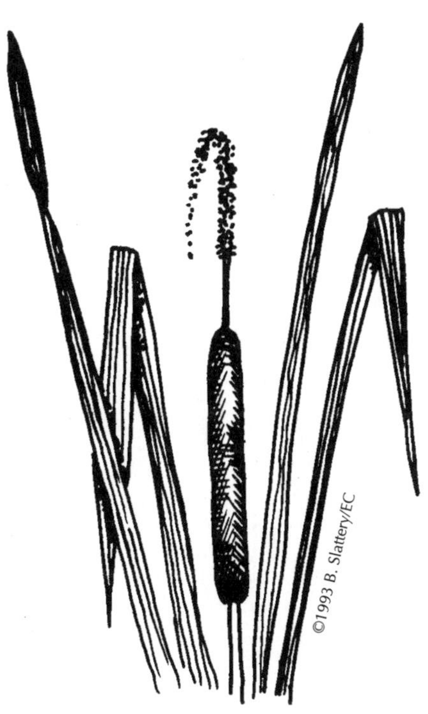

TYPHA ANGUSTIFOLIA

NARROW-LEAVED CATTAIL

Characteristics

- herbaceous, perennial, persistent, up to 10 feet in height
- rate of spread: rapid, > 1 foot per year by rhizome (in unconsolidated sediments)

Hydrology

- indicator status: obligate (OBL)
- salinity: up to 15 ppt
- tidal zone: upper 20% of the intertidal zone to spring tide elevation
- Nontidal regime: seasonally, irregularly, regularly, or permanently inundated to a depth of 1 foot, or saturated

Habitat

- brackish tidal marshes, fresh tidal marshes, nontidal marshes, shores of streams, ponds, and lakes

Notes

- requires full sun; soil stabilizer through root mat development; tolerates drought; pH preference is 3.7 to 8.5; considered an invasive species by some

Wildlife Benefits

- seeds, rhizomes: tule and snow geese, teal ducks, muskrats, beavers
- cover, nesting: canvasback duck, western grebe, marsh wren, red-winged blackbird, wood duck, gadwall, young fish

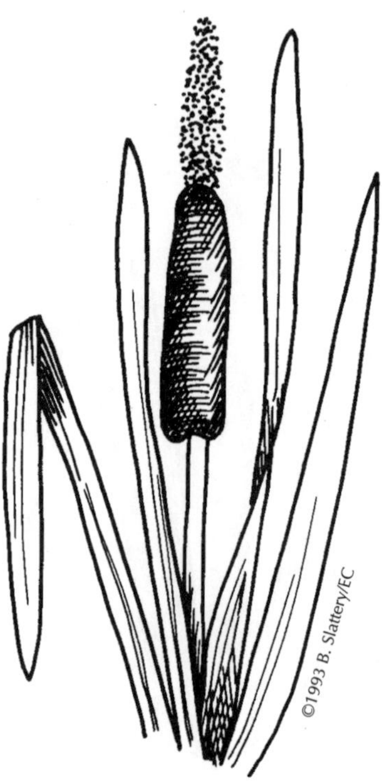

TYPHA LATIFOLIA

BROAD-LEAVED CATTAIL

Characteristics

- herbaceous, perennial, persistent, up to 10 feet in height
- rate of spread: rapid, > 1 foot per year (in unconsolidated sediments)

Hydrology

- indicator status: obligate (OBL)
- salinity: no tolerance
- tidal zone: upper 20% of the intertidal zone to spring tide elevation
- nontidal regime: seasonally, irregularly, regularly, or permanently inundated to a depth of 1 foot or saturated

Habitat

- fresh tidal marshes, nontidal marshes, shores of streams, ponds, and lakes

Notes

- requires full sun; soil stabilizer through root mat development; tolerates drought; pH preference is 3.7 to 8.5; considered an invasive species by some

Wildlife Benefits

- seeds, rhizomes: tule and snow geese, teal ducks, muskrats, beavers
- cover, nesting: canvasback duck, western grebe, marsh wren, red-winged blackbird, wood duck, gadwall, young fish

VACCINIUM CORYMBOSUM

HIGHBUSH BLUEBERRY

Characteristics

- deciduous shrub, 6 to 12 feet in height

Hydrology

- indicator status: facultative wet minus (FACW–)
- salinity: resistant, tolerates infrequent flooding by brackish water
- tidal zone: not found in tidal wetlands
- nontidal regime: seasonally inundated to a depth of 0.5 foot or saturated

Habitat

- forested wetlands, shrub swamps, bogs

Notes

- tolerates full shade; fairly insensitive to wind and ice damage; pH preference is 3.5 to 6.0 (will tolerate 6.5)

Wildlife Benefits

- fruit: blue jay, black-capped chickadee, tufted titmouse, brown thrasher, eastern bluebird, orchard oriole, pine grosbeak
- fruit, cover, and/or nesting: ruffed grouse, ring-necked pheasant, mourning dove, eastern kingbird, gray catbird, American robin, hermit thrush, rufous-sided towhee

VIBURNUM DENTATUM

SOUTHERN ARROWWOOD

Characteristics

- deciduous shrub, 6 to 12 feet in height
- rate of spread: slow, < 0.1 foot per year by suckers

Hydrology

- indicator status: facultative (FAC)
- salinity: resistant, tolerates infrequent flooding by brackish water
- tidal zone: above the spring tide elevation
- nontidal regime: seasonally inundated to a depth of approximately 0.5 foot or saturated

Habitat

- fresh tidal marshes, nontidal marshes, swamps, forested wetlands

Notes

- tolerates partial shade; male and female flowers on same plant; fairly insensitive to wind and ice damage; pH preference is 5.1 to 6.5

Wildlife Benefits

- fruit: common flicker, eastern phoebe, brown thrasher, American robin, eastern bluebird, white- and red-eyed vireos, rose-breasted grosbeak, pileated woodpecker, small mammals
- fruit, cover, and/or nesting: ruffed grouse, brown thrasher, gray catbird

VIBURNUM LENTAGO

NANNYBERRY

Characteristics

- deciduous tree, 20 to 35 feet in height

Hydrology

- indicator status: facultative (FAC)
- salinity: no tolerance
- tidal zone: not found in tidal wetlands
- nontidal regime: seasonally inundated to a depth of approximately 0.5 foot or saturated

Habitat

- forested wetlands

Notes

- tolerates full shade; fairly insensitive to wind and ice damage; pH preference is 6.1 to 7.5

Wildlife Benefits

- fruit: common flicker, American robin, eastern bluebird, cedar waxwing, rose-breasted grosbeak, purple finch, pileated woodpecker
- fruit and cover: ruffed grouse, bobwhite, ring-necked pheasant, hermit thrush
- fruit, cover, nesting: gray catbird

PART II

STUDENT ACTIVITIES

©2000 C. Barcomb/EC

8 DESIGNING A WETLAND

ACTIVITIES

Where Are We?

What Grows There?

Making Choices, Setting Goals

Are We Legal?

Finding Living Benchmarks

Catch a Flowing Stream

Will Stormwater Run Off?

Balancing Water Budgets

Where Are We?

SUMMARY

National Wetland Inventory (NWI) maps and topographic maps of the watershed in which the school is located will be read and interpreted. Potential sites and water sources for planned wetlands will be identified.

OBJECTIVES

Students will locate their school on appropriate topographic maps and National Wetland Inventory maps. Using topographic maps, students will determine the extent of the watershed surrounding the school and identify potential water sources for the planned wetland. On maps, they will locate wetlands near their school, then determine the types of wetlands using the NWI wetland classification system.

MATERIALS

- U.S. Geological Survey topographic maps of quadrangle containing the school
- National Wetland Inventory maps of quadrangle containing the school
- Copies of school site plan (check with Principal or Building Engineer)
- Student copies of topographic map area surrounding school
- Colored markers, pencils, or crayons
- Copies of *Maps* Student Pages

MAKING CONNECTIONS

To locate a water source for their planned wetland, students will identify the types of wetlands that surround the school and determine the location of the school relative to the watershed. As eventual home buyers, students will be able to use the same skills to determine the presence of wetlands on property and to site construction outside of wetland areas.

BACKGROUND

Read Chapter 1, especially the sections for *Hydrology, Planned Wetlands, Water Sources,* and *Common Wetland Types* (pages 7-21).

A watershed is the land area from which all water drains into a common body of water, such as a river, lake, or bay. Stormwater runoff follows the contours of the land as it flows from higher elevations to relatively low lying areas. If water flow is blocked, puddles, ponds, or lakes form. Flowing water forms streams, then rivers that lead to bays, then oceans. Watersheds can be as small as the school roof or as large as the Mississippi River drainage basin.

A topographic map, or "topo," is a useful tool to have on hand when exploring a watershed. It shows the relief (shape and elevation) of the land, as well as waterways and human-made features such as roads, buildings, and bridges. Contour lines show the elevations of ridges, hills, and other bumps and dips. Each contour line on a topo map represents an elevation in number of feet above sea level. Topo maps show locations of larger wetlands as well. Some helpful symbols used on topographic maps are shown in **Figure 8.1**.

Figure 8.1 These are frequently used topographic map symbols.

CONTOURS	BROWN	RIVERS & LAKES	BLUE
Depression	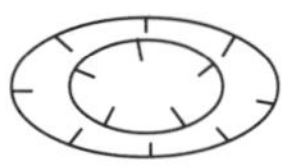	Intermittent stream	
Index contour		Intermittent river	
Intermediate contour		Disappearing stream	
SUBMERGED AREAS	**BLUE**	Perennial stream	
Marsh or swamp		Perennial river	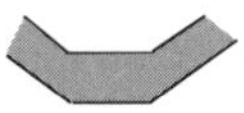
Submerged marsh or swamp	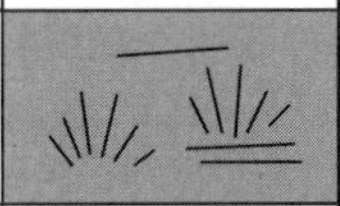	Intermittent lake or pond	
Wooded marsh or swamp	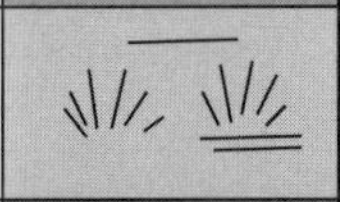	Dry lake	
Submerged wooded marsh or swamp	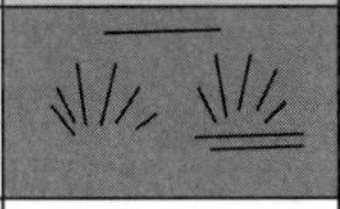	Narrow wash	
Land subject to inundation	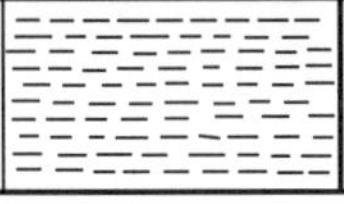	Wide wash	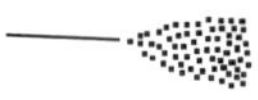

National Wetland Inventory maps, or NWI maps, correspond geographically to topo maps and employ the same quadrangle names (usually according to the largest community or most prominent land form on the sheet). Whereas topo maps provide information about the lay of the land (elevation lines), the primary information supplied by NWI maps is the location and identification of wetland areas. The Cowardin (1979) Classification System used on NWI maps is summarized in **Figure 8.2**. For subclasses and modifying terms examine the legend at the bottom of each NWI map.

Each of the five major wetland groups is subdivided into subsystems (except palustrine), classes, and subclasses, with additional modifying terms. Each wetland is described by a code that can be interpreted with the legend (key) on the NWI map. For example, a wetland with the code P-FO-1-H is a palustrine, forested, broad-leaved deciduous tree,

Figure 8.2 These descriptions are used in the NWI Map Classification System established by Cowardin et al. (1979). Additional classes have been added and some names have been changed.

SYSTEM	SUBSYSTEM	CLASS
M - MARINE 1. Open ocean overlying the continental shelf and associated coastline, 2. A salinity of greater than 30 ppt, and 3. Extends from the continental shelf inward to a. the extreme high tides, b. the seaward limit of wetland plants, c. the seaward limit of estuarine systems.	**1 - SUBTIDAL** Continuous inundation.	RB--Rock bottom
		UB--Unconsolidated bottom
		AB--Aquatic bed
		RF--Reef
		OW--Open water
	2 - INTERTIDAL Land exposed by tides.	AB--Aquatic bed
		RF--Reef
		FL--Flat
		RS--Rocky shore
		BB--Beach-bar
E - ESTUARINE 1. Deepwater tidal habitats and adjacent wetlands that are semi-enclosed by land, but have access to the ocean. 2. Ocean water occasionally diluted by fresh water, and 3. Limits: a. upstream to a minimum salinity of 0.5 ppt, b. the line crossing the entrance to a river, bay or sound, c. the seaward limit of wetland plants.	**1 - SUBTIDAL** Continuous inundation.	RB--Rock bottom
		UB--Unconsolidated bottom
		AB--Aquatic bed
		RF--Reef
		OW--Open water
	2 - INTERTIDAL Land exposed by tides.	AB--Aquatic bed
		RF--Reef
		FL--Flat
		SB--Streambed
		RS--Rocky shore
		BB--Beach-bar
		EM--Emergent
		SS--Scrub-shrub
		FO--Forested
L - LACUSTRINE 1. Wetlands and deepwater habitats in topographic depressions or dammed river channels, 2. Lacking trees, shrubs, persistent emergent vegetation, and/or lacking emergent moss/lichen with greater than 30% cover, 3. Areas greater than 8 hectares (20 acres), 4. Ocean-derived salinity of less than 0.5 ppt, 5. Limits: bounded by uplands or wetlands containing vegetation not described in #2.	**1 - LIMNETIC** All deep waters within this system.	RB--Rock bottom
		UB--Unconsolidated bottom
		AB--Aquatic bed
		OW--Open water
	2 - LITTORAL All wetland habitats within the lacustrine system from the shoreward boundary to a depth of 2 m (6.5 ft) or the maximum extent of nonpersistent emergent vegetation.	RB--Rock bottom
		UB--Unconsolidated bottom
		AB--Aquatic bed
		FL--Flat
		RS--Rocky shore
		BB--Beach-bar
		EM--Emergent
		OW--Open water

<table>
<tr><th>SYSTEM</th><th>SUBSYSTEM</th><th>CLASS</th></tr>
<tr><td rowspan="28">R - RIVERINE

1. All wetlands and deepwater habitats contained within a channel except:
a. wetlands dominated by trees, shrubs, emergent vegetation, and/or emergent mosses or lichens;
b. habitats with ocean derived salts greater than 0.5 ppt.
2. Limits:
a. bounded by an upland, channel bank, or wetland dominated by above-described vegetation;
b. braiding streams not included in this system.</td><td rowspan="9">1 - TIDAL

Water level fluctuates under tidal influence.</td><td>EM--Emergent wetland</td></tr>
<tr><td>RB--Rock bottom</td></tr>
<tr><td>UB--Unconsolidated bottom</td></tr>
<tr><td>AB--Aquatic bed</td></tr>
<tr><td>FL--Flat</td></tr>
<tr><td>SB--Streambed</td></tr>
<tr><td>RS--Rocky shore</td></tr>
<tr><td>BB--Beach-bar</td></tr>
<tr><td>OW--Open water</td></tr>
<tr><td rowspan="9">2 - LOWER PERENNIAL

Well-developed flood plain, has occasional water flow.</td><td>EM--Emergent wetland</td></tr>
<tr><td>RB--Rock bottom</td></tr>
<tr><td>UC--Unconsolidated bottom</td></tr>
<tr><td>AB--Aquatic bed</td></tr>
<tr><td>FL--Flat</td></tr>
<tr><td>SB--Streambed</td></tr>
<tr><td>RS--Rocky shore</td></tr>
<tr><td>BB--Beach-bar</td></tr>
<tr><td>OW--Open water</td></tr>
<tr><td rowspan="8">3 - UPPER PERENNIAL

High gradient, has fast-flowing water.</td><td>RB--Rock bottom</td></tr>
<tr><td>UB--Unconsolidated bottom</td></tr>
<tr><td>AB--Aquatic bed</td></tr>
<tr><td>FL--Flat</td></tr>
<tr><td>SB--Streambed</td></tr>
<tr><td>RS--Rocky shore</td></tr>
<tr><td>BB--Beach-bar</td></tr>
<tr><td>OW--Open water</td></tr>
<tr><td>4 - INTERMITTENT
Water flows only part of the year.</td><td>SB--Streambed</td></tr>
<tr><td colspan="2" rowspan="9">P - PALUSTRINE (No Subsystem)

1. All nontidal systems dominated by trees, shrubs, emergent vegetation, emergent moss/lichen, and all wetlands occurring in fresh tidal wetlands.
2. All areas lacking the above vegetation, but with all four of the following characteristics:
a. less than 8 hectares (20 acres) in size,
b. lacking a bedrock shoreline,
c. depth of the deepest part of the basin is less than 2 m (6.5 ft),
d. ocean-derived salts are less than 0.5 ppt.
3. Limits: bounded by uplands or any other of the systems previously described.</td><td>RB--Rock bottom</td></tr>
<tr><td>UB--Unconsolidated bottom</td></tr>
<tr><td>AB--Aquatic bed</td></tr>
<tr><td>FL--Flat</td></tr>
<tr><td>ML--Moss-lichen wetland</td></tr>
<tr><td>EM--Emergent</td></tr>
<tr><td>SS--Scrub-shrub</td></tr>
<tr><td>FO--Forested</td></tr>
<tr><td>OW--Open water</td></tr>
</table>

permanently flooded wetland. In other words, this is a freshwater forested wetland that has some surface water all year and a predominance of deciduous trees. The code R-4-SB-2 means riverine, intermittent, stream bed, sandy bottom. This is a stream with a sandy bottom that dries up during part of the year.

On NWI maps, always use the key associated with the appropriate system first (i.e., marine, estuarine, riverine, lacustrine, or palustrine). When the key is exhausted, move to the box containing modifiers in the lower right corner below the map.

PROCEDURE

Warm-Up

Ask students where the highest ground and the lowest ground are on school property. How do they know it is the highest (or lowest)? Which is wetter, the high ground or the low ground? Why?

Activity

Grades K-4

1. Go into the schoolyard. Find the spot that has the highest elevation (even if it is only slightly higher than the surrounding area). Each student can pretend to be a raindrop falling on this spot. All raindrops should slowly walk the path they might take as gravity pulls them downhill. Stress that water moves downwards to lower areas; it does not flow uphill. Students should stop anywhere they think the raindrop would stop. Some may not move at all, if they soaked (were absorbed) into the soil. Others may stop at a small dip or at the very bottom of a slope or hill to form a puddle. Still others may move around a stone or other object in their path. If there is a body of water nearby, some may wander all the way there. Afterwards ask students what caused them to take the path they chose, and what sorts of things might shape the path of flowing water, such as a stream.

2. Photocopy a topographic map in sections and accent major features such as roads (black) and waterways (blue) with colored highlighters; cut the map copy into pieces. These pieces may be laminated or covered in clear contact paper to protect them. Have students piece together the puzzle. Help them find the location of their school (see NOTE on next page), the watershed in which the school is located, and nearby wetlands. Use wet erase markers (transparency pens) to highlight these features.

3. Use a copier to make enlarged copies of the area of a topographic map showing the school location (see NOTE on next page). One copy will be needed for each group of students. If these enlargements are laminated or covered in clear contact paper, they become reusable. Using wet erase markers (transparency pens), students should find the school's location (circle in red), nearby waterways (color blue), wetlands (color green), roads (draw black), and the tops of hills and ridges (color red). Hills and ridges will look like irregular bull's-eyes.

Take these maps outside, and in a walk around school property, help students make the connection between the highest elevations of land

and the areas shaded in red on the map. Allow students to release one or more inflated balls from the higher elevations and observe where they stop rolling. Would these lower places be good locations for wetlands? Why?

Grades 5-12
NOTE: Schools and roads built after the topo and NWI maps were made will not appear on the maps. Determining the school's location on the maps should still be possible based on other features.

1. Before class begins, locate the school (or other site of interest) on the appropriate topographic map. Make copies of this area of the map for use by students. If possible, enlarge the area of interest.

2. Supply each student with a copy of the *Maps* Student Page. Complete Part I using one full-sized topographic map per group.

3. For Part II of *Maps* provide each student with a page-sized copy of the topo map area surrounding the school site. On these map copies students will circle the school, then mark the watershed surrounding the school. Remember that gravity causes rainwater to flow downhill, moving in a straight line until deflected by an object or an increase in elevation. Students will mark the direction in which rainwater flows from the surrounding higher elevations towards the school and where water flows away from the school to the local outlet for the watershed (a low elevation such as a swale, stream, pond, lake, river, bay, or ocean). See **Figure 8.1** for map symbols.

4. Using one full-sized National Wetland Inventory map per group, locate the school using roads and other familiar features. Locate three to five wetlands that are close to the school. Record the wetland codes and interpret the wetland types using the classification system at the bottom of the map. Record this information in Part III of *Maps*. **Figure 8.2** may by helpful, but it is not as complete as the classification at the bottom of the NWI map.

5. On the topographic map copies used in Part II, mark the wetlands within the watershed and those close to school. Is there a natural water source for a planned wetland? Complete Part IV of *Maps*.

Wrap-Up
Using copies of the school site plan, check elevations and contour lines to locate potential planned wetland sites. Compare with student watershed maps. Is there a natural source of water for a planned wetland? Are there designed water sources, such as parking lot runoff or swales? If so, select several potential sites (low spots) on the school grounds and mark them on the school site plan. Are there artificial sources of water for a planned wetland, such as downspouts or spigots? If so, note these locations on the school site plan also (use a different color).

ASSESSMENT
Have the *Maps* Student Pages been completed appropriately?
On student copies of topographic maps check the following:
- Is the school circled in red?

- Are the hills, peaks, and ridges shaded in red?
- Is the watershed containing the school site outlined?
- Is the water flow towards the school marked with green arrows?
- Is the water flow out of the watershed marked with blue arrows?

Name and describe the wetlands closest to the school. How many wetlands are within the watershed? Are there potential sites on school property for a planned wetland supplied by natural or artificial sources of water?

EXTENSIONS

Using maps, investigate other areas of the community to determine other potential sites for planned wetlands. Locate potential water sources for these planned wetlands. If service learning credits or scouting merit badges are being sought, students may wish to seek permission and funding to create a community wetland in a park or on other public property. Family property and community property values are often enhanced by planned wetlands.

Using math skills, figure the percentage of wetland area within the watershed. One technique for measuring an irregular area is described on page 164.

RESOURCES

Cowardin, L.M., V. Carter, F.C. Golet, and E.T. LaRoe. 1979. *Classification of Wetlands and Deepwater Habitats of the United States.* U.S. Fish and Wildlife Service, Washington, DC. FWS/OBS-79/31.

Dahl, T.E. 1990. *Report to Congress: Wetland Losses in the United States 1780's to 1980's.* U.S. Department of the Interior, Washington, DC.

Dahl, T.E. and C.E. Johnson. 1991. *Status and Trends of Wetlands in the Conterminous United States, Mid-1970's to Mid-1980's.* U.S. Fish and Wildlife Service, Shepardstown, WV.

Kesselheim, A.S. and B.E. Slattery. 1995. *WOW!: The Wonders of Wetlands.* Environmental Concern Inc., St. Michaels, MD.

Smith, G.S. 1991. *NWI Maps Made Easy: A User's Guide to National Wetlands Inventory Maps of the Northeast Region.* U.S. Fish and Wildlife Service, Hadley, MA.

Tiner, R. "NWI Maps–Basic Information on the Nation's Wetlands" In *BioScience.* May 1997.

To obtain topographic maps:
U.S. Geological Survey, 509 National Center, Reston, VA 20192
1-800-USA-MAPS or www.usgs.gov
Downloadable digital maps are available at:
www.terraserver.microsoft.com
Local map distributors are listed at:
mapping.usgs.gov/esic/map_dealers

To obtain National Wetland Inventory maps:
USGS/ESIC, 507 National Center, Reston, VA 22092
1-888-ASK-USGS or www.nwi.fws.gov/order_maps.htm
Downloadable digital maps are available at: www.fws.gov
(click on Wetlands, then Wetlands Interactive Mapper)

MAPS

Part I: Topographic Maps

1. Name the quadrangle shown on the map, the date the map was drawn, and the dates of any revisions (in purple).

2. What is the increment between (brown) contour lines?

3. The symbol for a school is a small square with a flag on top. Locate your school on the topo map. What is the elevation of your school?

4. Bodies of water are colored blue. Name three bodies of water near your school.

5. What does this blue symbol represent?

6. What is the elevation of the wetland nearest your school?

7. What are the elevations of the three highest peaks or hills near your school?

Part II: Watersheds

Locate the school site on your copy of the topo map; the symbol is a small square with a flag on top. Mark the school by drawing a small red circle around it.

Hill tops will be closed irregular circles, and ridge tops will be closed elongated shapes. The highest peaks will look like bull's-eyes and may have a brown + in the middle with a number indicating the elevation. Locate the hill tops and ridge tops surrounding the school site; color them red. Remember that enclosed blue-shaded areas are bodies of water such as ponds and lakes. Depressions are closed shapes with perpendicular lines pointing inwards.

Connect the red-shaded areas surrounding the school site. This is the primary watershed for the school.

Gravity causes water to flow downhill, moving in a straight line until deflected by an object or an increase in elevation. Mark with green arrows the downward flow of rainwater from each peak, hilltop, and ridge towards the school. Carefully examine the elevations within the watershed. Mark with blue arrows where water flows away from the school and out of the watershed.

Part III: National Wetland Inventory Maps

In the first column of the table below, list the map codes for wetlands within the watershed in which the school is located. If no wetlands are shown within the watershed, then list the codes for those wetlands nearest the school.

Use the classification at the bottom of the NWI map to decipher the wetland codes that have been entered in the table below. Remember to stay within the appropriate system (such as Palustrine) until you run out of choices, then go to the box in the lower right corner to determine the water regime and other modifiers.

Map Code	System	Subsystem	Class	Subclass	Water Regime	Other Modifiers

Part IV: School Site Plan

8. Name the possible natural water sources for your planned wetland.

9. Name the possible artificial water sources for your planned wetland.

What Grows There?

SUMMARY

Students assess soils and plants along transects at the planned wetland site to determine whether wetland characteristics are present and to provide baseline data for future decisions concerning the site. Hydrology will also be assessed for wetland characteristics.

OBJECTIVES

Data on soils and plants will be collected along transects through the planned wetland site. Soil color and texture will be assessed and evaluated for wetland characteristics. Plant forms will be determined, plant species identified, and frequency of occurrence in wetlands assessed. The site will be examined to determine if wetland hydrology is present. Plants, soils, and hydrology will be evaluated to determine if a wetland ecosystem exists.

MATERIALS

- An enlarged site map or drawing of the area to be assessed
- Compass
- Measuring tape (100 feet)
- Meter sticks
- String to mark transects
- Clipboards and pencils
- Plant identification guide books
- Soil probes or soil shovels
- Jug of water
- Soil color charts (commercial or *Color Me Wet!* in *WOW!: The Wonders of Wetlands)*
- Copies of *Plant Transect* Student Page
- Copies of *Key to Soil Texture* Student Page
- Copies of *Soil Transect* Student Page
- Copies of *Wetland Characteristics* Student Page

MAKING CONNECTIONS

To make informed decisions at any time in our lives, appropriate data is needed. Before planning a wetland, baseline information on vegetation, soils, and hydrology should be collected and analyzed. This provides information about whether the site has been a wetland in the past, has wetland characteristics currently, or has potential for becoming a wetland site. Informed decisions in wetland planning can help us prevent damage to functioning wetlands, and the waste of time, money, and effort to prepare and plant an area that cannot function as a wetland.

BACKGROUND

Read Chapter 1, especially sections on *Vegetation*, *Soils*, and *Hydrology* (pages 5-7).

A wetland is the boundary between land and water, with some features of each and with some features that are uniquely different. Three characteristics are used in delineating wetlands: hydrologic regime, hydric soils, and hydrophytic plants. The presence of surface water or groundwater within the top eighteen inches of soil for approximately two weeks during the growing season is an indication of wetland hydrology.

If wetland hydrology is or has been present, typically it will cause soils to become gleyed where they are continually wetted (i.e. show subdued coloration from frequent or repeated wetting, like a colored T-shirt after repeated washing). Wetland soils may contain oxidized rhizospheres where oxygen leaks from plant rootlets. Wetland hydrology will produce mottling (red or black specks) where the soil is alternately wetted and dried, or have a deep organic layer due to slower decomposition rates under anaerobic conditions. Clay-based soils hold water more readily than sand-based soils, wick the water higher above the water table, and extend the hydric conditions closer to the soil surface. Silt-based (loam) soils also hold water, but not as well as clay.

In the presence of wetland hydrology and hydric soils, hydrophytic plants are uniquely adapted for survival in wet conditions. Most adaptations bring plant roots into contact with oxygen from the air, either by bringing roots to the air or air to the roots.

Several activities in *WOW! The Wonders of Wetlands* (1995) are helpful in developing skills for identifying hydric soils and hydrophytic plants. It is recommended that these activities from *WOW!* be experienced independently, before being integrated into this one activity:

This Plant Key Is All Wet! (pages 123-128)
Tracking Plants and Keeping Track (pages 138-142)
Do You Dig Wetland Soil? (pages 231-238)

PROCEDURE

Warm-Up

Sometimes wetlands do not seem wet at all. Encourage students to give examples of areas that are sometimes wet or have been wet in the past (puddles, ditches, drained and filled wetlands, vernal pools).

Introduce the three factors considered when determining whether an area is a wetland (soils, plants, hydrology), and discuss how they might be connected.

Activity

Grades K-4

Do as much of the activity listed for grades 5–12 as students can manage, using the worksheets provided when appropriate.

A. PLANTS
Look carefully around the planned wetland site. Are emergent or aquatic plants present? Are they found all over the site or just in one area? Are they all of the same kind (such as cattails)?

B. SOILS
Use the soil probes or soil shovels to examine the soil from the surface

to eighteen inches deep at the planned wetland site. Compare the soil colors with those of a soil chart. Could the soil be wetland soil? What is the texture of the soil? The presence of a large proportion of clay in the soil means it is more likely to hold water.

C. HYDROLOGY
Does the hole made with the soil probe or shovel fill with water? Are dried mud cracks present in low spots? Are plants water-marked or mud-stained, showing how high the water has been? Are there low spots where water could or does collect? Are signs of water erosion present, such as gullies and channels?

Grades 5-12
A. ESTABLISHING TRANSECTS
1. Choose a prominent object or feature near the wetland site from which compass bearings and distances can be measured to each transect. Mark and describe this benchmark object or feature on the site map or drawing of the area. If a tree, rock, sidewalk, or other semipermanent feature is not available, create one with a wooden stake and spray paint or plastic ribbon.

2. Divide the group into three, four, or five teams. Each team will be assigned a letter label and will collect data along one transect.

3. Place teams at three-meter (or ten-foot) intervals along one side of the wetland site. Mark the starting point for each team with a survey flag or stake to which the ball of string will be attached. Have each team unroll their string across the wetland site parallel to the other teams. The string marks the transect line for each team.

4. Measure the distance and compass bearing from the benchmark to the starting point of each transect. Mark both the distance and compass bearing on the site map or drawing of the planned wetland site, then mark the transect lines on the map and label them with the appropriate transect letters. See **Figure 8.3a.**

B. PLANTS
1. Review the following definitions of types of plants:

- Tree–usually a single woody rigid trunk; more than 20 feet tall
- Shrub–multiple stems that are woody and rigid; less than 20 feet tall
- Vine–stems may be woody and rigid, but leaning on other plants
- Emergent–pliable stems; growing in wet soil or water with part of stem and leaves above water
- Aquatic–pliable stems; growing completely under water, floating on the surface, or with floating leaves

2. Aquatic and emergent plants are adapted to wetland conditions. Some trees, shrubs, vines, and grasses are also adapted to wetlands. Hydrophytic plants respond to the stimulus of prolonged soil moisture in some of the following ways:

- Shallow roots--Since water is readily available, roots need not grow deep. Oxygen is in short supply and is near the surface.

- Adventitious roots--These roots can be seen at or above the soil surface; they seem to assist in obtaining oxygen.
- Shoot elongation--During high water events, fast shoot growth can keep stems and leaves out of the water so photosynthesis and movement of oxygen to the roots may continue.
- Hypertrophied (enlarged) lenticels or their scars may be seen on woody wetland plants. Lenticels permit gas exchange through bark and enlarge during floods.

Wetland plants have adaptations that allow them to live in wet conditions. Some adaptations that are readily visible are:

- Aerenchyma--This tissue of enlarged cells (readily visible in cattail stems) moves oxygen downward to the roots.
- Hollow stems--Most easily seen in reeds, this adaptation moves oxygen downward to the roots.
- Buoyant fruits and seeds--Many wetland plants use water to disperse their progeny.

3. Provide each team with copies of the *Plant Transect* Student Page (where they will record the letter label for their transect), a ball of string, and a meter stick or tape measure. Beginning with the survey flag or stake marking the beginning of each transect, measure three meters (or ten feet) along the transect, and mark or knot the string. This is Section #1. Tally each type of plant seen along this section of the transect. Tally all plants touching the string, hanging over the string, under the string, or within one foot of the string on both sides of the string. Identify the most common plants found along this section of the transect.

4. Measure another three meters (or ten feet) along the transect; mark or knot the string. This is Section #2. Tally the plants for this section, again identifying the most common plants. Repeat until the team has crossed the wetland site. Leave the transect lines in place for use in part C.

5. Optional. Determine the wetland indicator status of the most common plants identified along the transect. This information is available in Chapter 7 and at the USDA plant data website (see Resources).

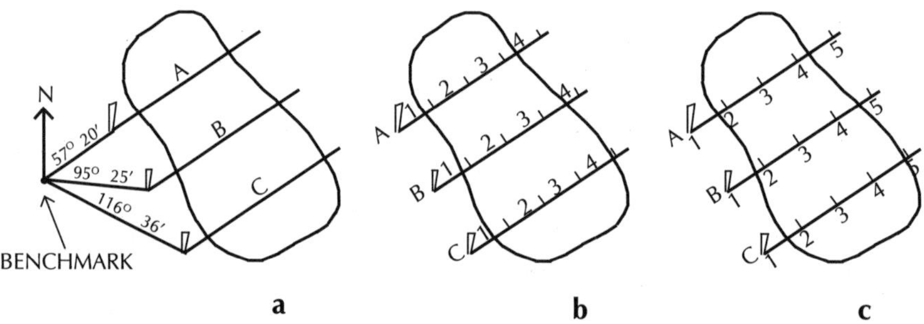

Figure 8.3 (a) Locate a transect benchmark, then measure distance and bearing to each transect. (b) Section locations along plant and hydrology transects. (c) Sample stations along soil transects.

C. SOILS
1. Provide each team with a soil color chart, the Key to Soil Texture, water, a soil probe or soil shovel, and a copy of the *Soil Transect* Student Page.

2. Following the same transect as in part A, each team should examine one soil sample at the beginning of Section #1, at the beginning of each succeeding section, and at the end of the last section. See **Figure 8.3c**. Examine the soil moisture, color, texture, and other features at depths of 6 inches (15 cm), 12 inches (30 cm), and 18 inches (45 cm). This data should be recorded on the *Soil Transect* Student Page. (To sample soil 18 inches below the surface, insert the probe into the hole formed when the shallower soil sample was collected.)

In the Northeast, soils are considered hydric if a peat or muck surface layer is $\geq$ 8 inches deep, or if the matrix color has a chroma $\leq$ 2 with mottles, or if the matrix color has a chroma of $\leq$ 1 without mottles. Check with the local Soil Conservation Service for information on what constitutes a hydric soil in your area.

D. HYDROLOGY
1. Observe the hydrology of the planned wetland site. Is standing water or tidal water present? Does water collect in a hole 18 inches deep (the zone where most plant roots are located)? Do other signs indicate that water has been present during the growing season?

2. Complete the *Wetland Characteristics* Student Page for each transect. Under **Yes**, **Maybe**, and **No**, record the sections of the transect to which each answer applies.

Wrap-Up

Students will evaluate the planned wetland site according to the following criteria:
a. Are wetland plants present? Yes, maybe, no.
b. Are wetland soils present? Yes, maybe, no.
c. Is water present during the growing season? Yes, maybe, no.

On the site map or drawing, students will label each section of their transect with their evaluations: green Y (yes), M (maybe), or N (no) for vegetation; brown Y, M, or N for soils; and blue Y, M, or N for hydrology.

ASSESSMENT

Have students evaluate whether all, some, or none of the area assessed is a wetland, and give reasons for their evaluation. This may be done orally as one large group, within teams and then reported to the larger group (orally or in writing), or individually in writing.

- If the area is a wetland, do students think it is healthy?
- Does it have a variety of plants, or are there large patches (the size of a table) of the same kind of plant?
- Do some areas show signs of erosion from stormwater runoff?
- Could silt be seen on the leaves of the plants in the wetland?
- If the area is not now a wetland, could it become one?
- Would the soils present hold water?
- Is there a source of water nearby?

EXTENSIONS

Investigate further and suggest how this site could be improved for wildlife. Assess and compare other possible wetland sites.

RESOURCES

Environmental Science: Water and Air. 1995. Globe Fearon Education Publisher, Paramus, NJ. [Especially: How do different soils affect the rate of absorption?]

Kesselheim, A.S. and B.E. Slattery 1995. *WOW!: The Wonders of Wetlands*. Environmental Concern Inc., St. Michaels, MD.

Lewis, W.M. Jr., Chair. 1995. *Wetlands: Characteristics and Boundaries*. National Academy Press, Washington, DC.

Tiner, R.W. 1998. *In Search of Swampland: A Wetland Sourcebook and Field Guide*. Rutgers University Press, New Brunswick, NJ.

On-line resources:

Wetland indicator status, information, and pictures for plants in all regions of the U. S. are located at: plants.usda.gov/plants/index.html.

Common persimmon, *Diospyros virginiana*

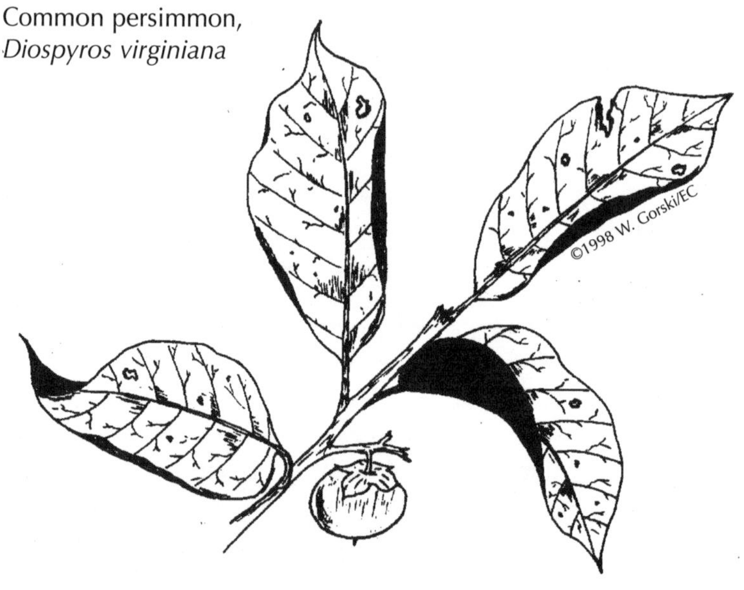

PLANT TRANSECT

Transect ____

Type of Plant	Species	Wetland Indicator Status	Number of Plants Counted					
Section Number:			1	2	3	4	5	6
Trees								
Shrubs								
Vines or grasses								
Emergent plants								
Aquatic plants								

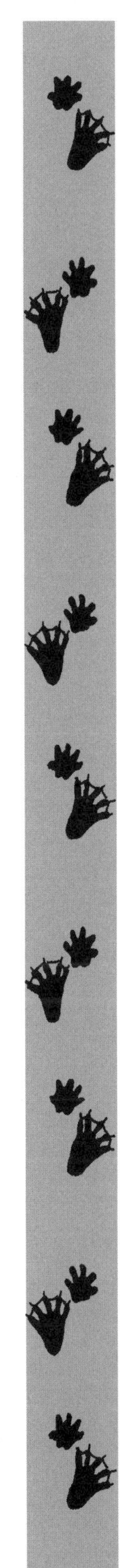

KEY TO SOIL TEXTURE

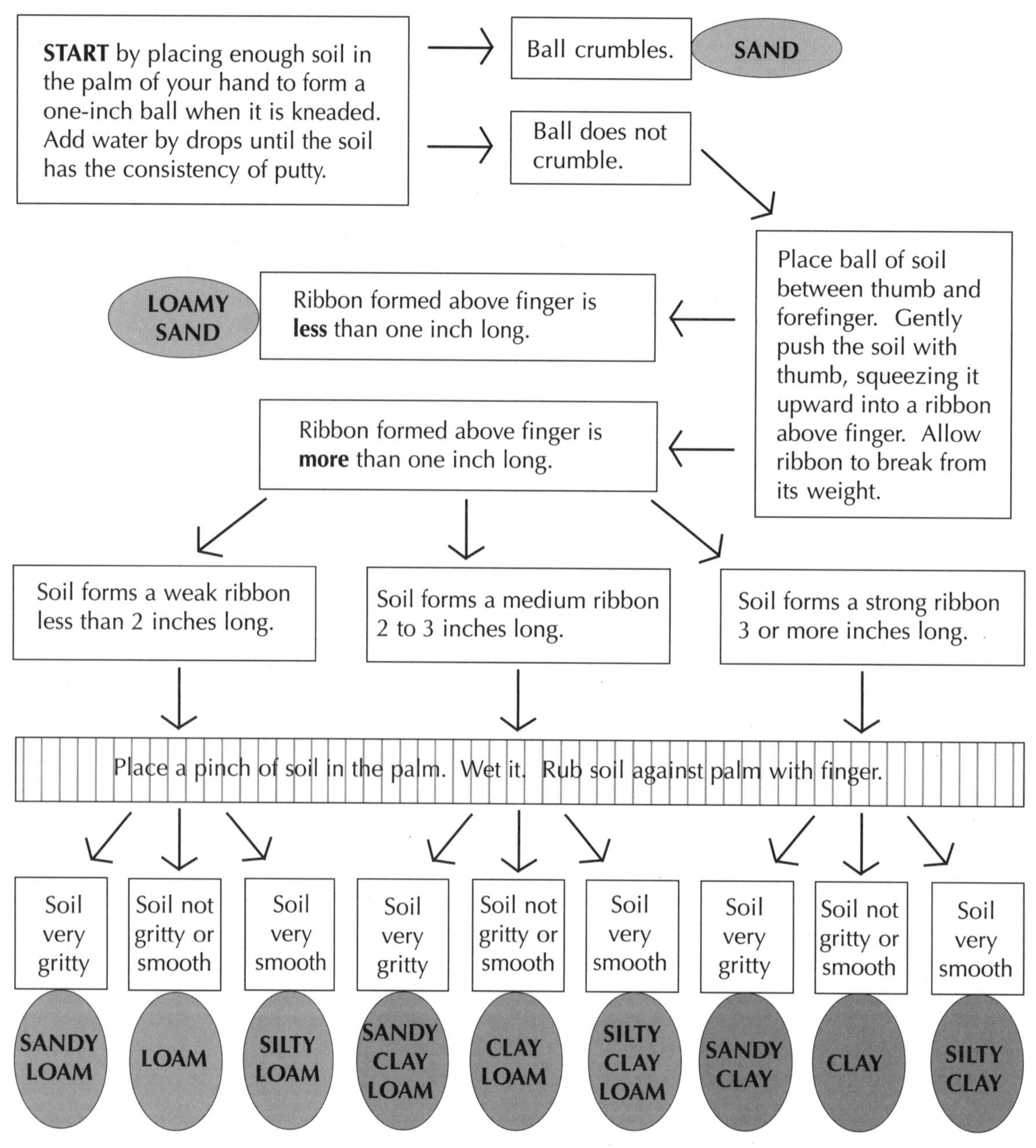

Adapted from *WOW!: The Wonders Of Wetlands* (1991) with permission.

SOIL TRANSECT

Transect _____

Record the words or phrases that apply to each soil sample in the chart below.
Moisture: Indicate whether dry, moist, wet, very wet, or drippy.
Texture: Designate as sand, loamy sand, sandy loam, loam, silty loam, sandy clay loam, clay loam, silty clay loam, sandy clay, clay, silty clay.
Color: Record number if using *Color Me Wet!*; record hue/value/chroma if using commercial soil color chart.
Other: Note pebbles, organic matter, rotten egg smell, mottling, oxidized rhizospheres.

Depth of Soil from Surface	Feature	Station 1	Station 2	Station 3	Station 4	Station 5	Station 6	Station 7
6 inches (15 cm)	moisture							
	texture							
	color							
	other							
12 inches (30 cm)	moisture							
	texture							
	color							
	other							
18 inches (45 cm)	moisture							
	texture							
	color							
	other							

WETLAND CHARACTERISTICS

Transect _____

Record the sections of the transect, or the stations along the transect, to which each statement applies.	**Yes**	**Maybe**	**No**
A. HYDROLOGY			
1. Is standing water or tidal water present?			
2. Does water collect in a hole 18 inches deep?			
3. Are there depressions where water might collect?			
4. Do low spots have mud or dried mud cracks?			
5. Do tree trunks and other vegetation appear water-stained?			
6. Is vegetation mud-stained from previous flooding?			
7. Are gullies, stream channels, or signs of water erosion present?			
B. SOILS			
1. Are soils wet?			
2. Excluding the surface organic layer, are soil colors those of hydric soils (green, dark gray, brown, black)?			
3. Are soil colors gleyed?			
4. Do soil samples show mottling or contain oxidized rhizospheres?			
5. Do soil samples have a high percentage of clay?			
6. Do the soils have a rotten egg smell?			
C. PLANTS			
1. Are aquatic or emergent plants present?			
2. Are any of the common plants typically found in wetlands?			
3. Are obligate wetland plants present?			

D. SUMMARY FOR THIS TRANSECT

1. Along which sections are wetland plants present? ____________________

2. At which stations are wetland soils present? ____________________

3. Which sections have water present during the growing season? ____________________

Making Choices, Setting Goals

SUMMARY

Wetland functions will be selected for incorporation in the planned wetland design. Goals for the wetland project will be established.

OBJECTIVES

Students will compare wetland functions and values, and select those to be incorporated in the planned wetland. Students will formulate goals for the project as they model decision-making skills.

MATERIALS

- Copies of *Choices* Student Page
- Overhead transparencies of *Group Choices* (optional)
- Overhead projector (optional)
- Copies of *Wetland Goals* Student Page

MAKING CONNECTIONS

Not all wetland functions are compatible, nor are all goals financially attainable. As with most decisions in our lives, choices must be made. Students have an opportunity to promote their personal interests in the planned wetland, as group goals are set for the wetland design.

BACKGROUND

Read Chapter 1, especially *Wetland Functions and Values* (pages 21-24).

A wetland function is a task performed by a wetland regardless of how society values that task. The structure of a wetland determines which functions it can perform. If the wetland is to function as fish habitat, then it must be deep enough for the species of fish desired. If the wetland is to function as frog habitat, it needs gently sloping sides for young frogs to climb out.

A wetlands's value is subjective; it is the worth society places on the attributes of a the wetland and may change over time. For instance, one wetland function is to provide a nursery for many creatures. Society sees this as valuable if the creatures are fish, frogs, and birds—but NOT valuable if biting insects like mosquitoes breed there.

One choice that must be made is whether the planned wetland will be created from a non-wetland area, or an existing wetland will be enhanced, or an area that was previously a wetland will be restored to wetland status.

Created wetlands result from modification of a site where wetlands do not currently exist and where wetlands did not previously exist. Beavers create wetlands by flooding areas that were previously dry. At schoolyard sites, conditions can be created that cause a wetland to exist. This generally means providing water to a planned wetland site and ensuring that sufficient water will remain at the site to provide wetland conditions, as well as any other conditions needed to support the wetland community. This can be accomplished through several means, but generally includes excavation to lower the ground elevation and construction of an earthen dike or berm to impound water. Sometimes a liner of plastic or clay is needed. It is important to understand that just digging a hole does not create a wetland.

Wetland restoration means changing an altered wetland from its present impacted condition back to that of a functioning wetland. Restoration projects can be classified as hydrologic or biological. Hydrologic restoration usually involves removing barriers that block water flow to a site (as in some tidal wetland areas), or plugging tile drains or drainage ditches to restore preexisting water regimes (as in the Prairie Pothole Region of the Midwest and the swamp areas of the Eastern Coastal Plain). Although biological restoration generally means reestablishing wetland plant species along with hydrologic restoration, in some cases restoring the biological portion of a system may be all that is needed. Some marginal agricultural lands, which were at one time wetlands, have been successfully replanted with wetland tree saplings in an effort to increase the acreage of bottomland hardwood forests in the South.

Wetland enhancement generally is a process by which some wetland functions are improved as part of an overall management plan. In the process, other wetland functions may be minimized. Wetlands may be enhanced by creating habitat for rare or endangered species through construction, placement, and maintenance of nesting structures. This may, however, cause a decline in other wetland species through predation or competition. A site dominated by invasive or alien plant species might be replanted with a variety of more desirable native species that support native wildlife. Wetland water levels are often managed (some seasons high, some seasons low) to support a variety of waterfowl in wildlife management areas.

After choosing the type of planned wetland and identifying a potential wetland site, appropriate specific goals for the project should be set. For example, if the site chosen is a low area in the topography with predominantly silt/clay soil and water present only during the spring, then a goal of providing habitat for frogs, turtles, butterflies, and birds would be appropriate, but fish habitat would not be possible.

PROCEDURE

Warm-Up

A function is a job that is performed or a role that is filled. A value is the degree of importance (either positive or negative) that is associated with a function. Discuss the difference between a function and a value. Provide an example (such as the function and value of butterflies or cows), then have students suggest some wetland functions. As students offer suggestions, list them on the board.

Activity

<u>Grades K-4</u>

Using an overhead projector, do the following activity as a class. As an alternative, have students in higher grades interview younger students about what they want the wetland to accomplish.

Grades 5-12

1. Each student should consider the list provided on the *Choices* Student Page, placing a check under **My Choice** for eight wetland functions that they would like the planned wetland to perform. Add any that do not appear on the list. Do not limit possibilities even if some seem contradictory; this can be sorted out later.

2. Consult other teachers and other students, the administration, those in charge of maintenance, the parent-teacher organization, local garden clubs, etc. Listen to concerns as well as desires. If the school administration says "no open water," make sure that appears on the list even if it is a negative characteristic.

3. Within small groups, discuss the appropriateness of each function, their importance, and their compatibility. Within each group select up to eight functions desired for the planned wetland. Mark these choices on the *Choices* Student Page under **Group Choice**.

4. Report back to the larger class each group's goals for the planned wetland. Compare lists by writing them on the board, on an overhead transparency of *Group Choices*, or on large (poster-size) sheets of paper that can be temporarily posted along a wall. Which goals are on most lists? Which are on only one? Why? Which items are too expensive? Consider postponing them until a later date when more funds may be available. Which items are inconsistent with the amount and timing of water available within the planned wetland? These must be eliminated unless an additional supply of water is available.

5. As a class, select a final list of wetland functions that are consistent with the water supply, administrative parameters, and available funds. Mark these in the column **Class Choice** on the *Choices* Student Page.

Wrap-Up

Will you be creating, restoring, or enhancing a wetland? Record the type of planned wetland to be designed on the *Wetland Goals* Student Page.

Using information from **Figure 8.2** and the NWI maps, determine the type of wetland to be designed. Record this on the *Wetland Goals* Student Page.

Circle the water sources available and the potential water losses on *Wetland Goals*.

Now record the wetland functions and any special characteristics selected by the class as goals. Keep these goals in mind as the planned wetland is designed. If unsure how to attain each goal, then individuals or small groups should investigate and report back to the class.

ASSESSMENT

Did each student participate in the decision-making process? Are the wetland goals compatible? Is the wetland project reasonable in view of the muscle power and financial resources available?

EXTENSIONS

Visit a nearby planned wetland, such as a stormwater runoff pond, and identify the functions of that wetland. This can also be accomplished with pictures or videos of wetlands.

RESOURCES

Mitsch, W.J. and J.G. Gosselink. 1986. *Wetlands.* Van Nostrand Reinhold Co., New York, NY.

Salvesen, David. 1990. *Wetlands; Mitigating and Regulating Development Impacts.* ULI–the Urban Land Institute, Washington, DC.

Sather, J.H. and R.D. Smith. 1984. *An Overview of Major Wetland Functions and Values.* U.S. Fish and Wildlife Service, Washington, DC.

Smith, R.D., A. Ammann, C. Bartoldus, and M.M. Brinson. 1995. *An Approach for Assessing Wetland Functions Using Hydrogeomorphic Classification, Reference Wetlands, and Functional Indices.* Technical Report WRP-DE-9. U.S. Army Corps of Engineers, Vicksburg, MS.

Tiner, R.W., Jr. 1984. *Wetlands of the United States: Current Status and Recent Trends.* U.S. Fish and Wildlife Service, National Wetlands Inventory, Springfield, VA (#PB90-198201).

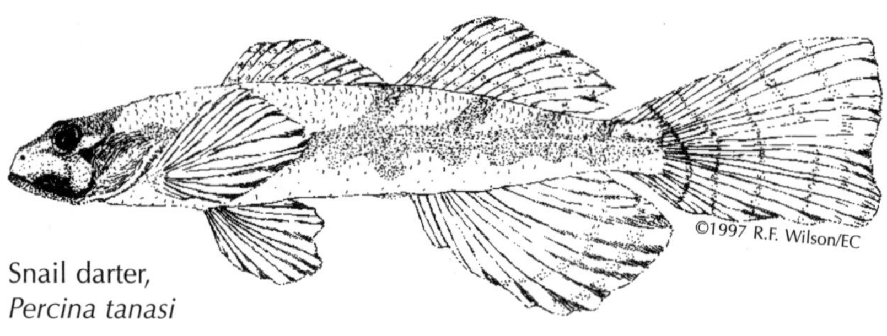

Snail darter,
Percina tanasi

CHOICES

In the first column, place a check mark next to the functions and values that are important to you as an individual. After discussion, list the group choices. Record class choices when they are made.

WETLAND FUNCTIONS:	**MY CHOICE**	**GROUP CHOICE**	**CLASS CHOICE**
Shoreline erosion control	____	____	____
Sediment stabilization	____	____	____
Water quality	____	____	____
Pollution abatement	____	____	____
Physical filter of impurities	____	____	____
Flood water storage	____	____	____
Groundwater recharge	____	____	____
Mammal habitat	____	____	____
Bird habitat	____	____	____
Reptile habitat (turtles)	____	____	____
Amphibian habitat (frogs)	____	____	____
Fish habitat	____	____	____
Butterfly habitat	____	____	____
Aquaculture (fish, crayfish, clams)	____	____	____
Food production (rice, cranberries)	____	____	____
Aesthetics (beauty)	____	____	____
Active recreation (fishing, ice skating)	____	____	____
Passive recreation (relaxation, bird watching)	____	____	____
Open space buffer	____	____	____
Educational activities	____	____	____
Research	____	____	____
____________________	____	____	____
____________________	____	____	____
____________________	____	____	____
____________________	____	____	____
____________________	____	____	____
____________________	____	____	____
____________________	____	____	____
____________________	____	____	____
____________________	____	____	____

GROUP CHOICES

After listing choices for each group, circle or highlight the functions selected by the class.

Wetland Functions:	**GROUPS** 1	2	3	4	5	6
Shoreline erosion control	___	___	___	___	___	___
Sediment stabilization	___	___	___	___	___	___
Water quality	___	___	___	___	___	___
Pollution abatement	___	___	___	___	___	___
Physical filter of impurities	___	___	___	___	___	___
Flood water storage	___	___	___	___	___	___
Groundwater recharge	___	___	___	___	___	___
Mammal habitat	___	___	___	___	___	___
Bird habitat	___	___	___	___	___	___
Reptile habitat	___	___	___	___	___	___
Amphibian habitat	___	___	___	___	___	___
Fish habitat	___	___	___	___	___	___
Butterfly habitat	___	___	___	___	___	___
Aquaculture	___	___	___	___	___	___
Food production	___	___	___	___	___	___
Aesthetics	___	___	___	___	___	___
Active recreation	___	___	___	___	___	___
Passive recreation	___	___	___	___	___	___
Open space buffer	___	___	___	___	___	___
Educational activities	___	___	___	___	___	___
Research	___	___	___	___	___	___
______________	___	___	___	___	___	___
______________	___	___	___	___	___	___
______________	___	___	___	___	___	___
______________	___	___	___	___	___	___
______________	___	___	___	___	___	___
______________	___	___	___	___	___	___
______________	___	___	___	___	___	___
______________	___	___	___	___	___	___

WETLAND GOALS

Circle the choices that are appropriate and fill in the blanks.

Our planned wetland: created, restored, enhanced

The type of planned wetland:

system ______________________

subsystem ___________________

class _________________________

Water sources: precipitation, surface water inflow, groundwater inflow

Water losses: evapotranspiration, surface water outflow, groundwater outflow

Our goals:

____

____

____

____

____

____

____

____

____

____

Are We Legal?

SUMMARY

Wetland hydrology, wetland soils, and hydrophytic plants indicate that an area may be a wetland. Federal, state, and local regulations use these indicators to recognize and control activities in wetlands, and to protect wetlands from destruction. Some wetland activities require permits, some do not, and some are prohibited. A permit may or may not be necessary for a planned wetland.

OBJECTIVES

Students will determine whether a permit is needed for their planned wetland.

MATERIALS

- Student scissors
- Copies of *Rules* Student Page
- Copies of *Permits* Student Page

MAKING CONNECTIONS

Many land use activities require a permit before action can be taken. Construction permits, building permits, and occupancy permits are examples of permits for which students may need to apply sometime during their lives. Rules and regulations exist to protect people and resources, and to draw the line between individual and group rights.

BACKGROUND

Read Chapter 3, *Permits* (pages 47-53).

In the past, destruction of wetlands was encouraged because they were considered to be foul-smelling places that harbored unsavory creatures and biting insects. Dredging, filling, paving, riprapping, and bulkheading occurred frequently. Wetlands were considered undesirable places, so the land was cheap.

In the 1960s and 1970s, people began to understand the importance of wetlands to the quality of our lives, and began to change behaviors to fit the new understanding of wetlands. Attitudes and behaviors, however, are slow to change. Over half the wetlands in the United States that were in existence when the first European settlers arrived have been destroyed. The wetland regulations that we now have may seem daunting at times, but their purpose is to prevent the destruction of the wetlands that remain, while people learn to understand and appreciate their value as a natural resource.

Rules and regulations prevent us from destroying our wetlands by draining, dredging, filling, paving, or bulkheading them without careful consideration of alternatives. If the decision is made to allow the destruction of a wetland, then it must be replaced with a larger one nearby. This is called mitigation. Destroying a wetland might create more farmland for a farmer, a neater yard for a waterfront homeowner, or more parking for a waterfront restaurant, but there will be far fewer water birds, fish, and shellfish for the rest of us to observe, eat, and enjoy. Also, the water that many of us drink will not be as clean, and floods will cause more extensive damage (even some distance inland behind the waterfront properties).

PROCEDURE

Warm-Up

Name several actions for which there are rules and regulations, or for which a permit or license is required. [Classroom rules, school rules, driving permits, etc.] What would happen if there were no rules and regulations? Drivers could drive anywhere and ignore stop signs, but would that be in the best interests of the rest of the people that share the road? Would you want to be in a school bus on the same road with them? Sometimes it is necessary to give up some personal freedom for the good of the group. Rules, regulations, and permits are a way of limiting personal freedoms for the benefit of the public as a whole. This is also true in wetlands. Eliminating wetlands on waterfront property may seem good for a landowner building a home, but it affects all neighboring properties and the larger ecosystem.

Activity

Grades K-4

Draw a happy face and a sad face on the *Rules* Student Page. Cut out the faces and each action. Place each action in a happy face pile or a sad face pile. In small groups have students share what they have in each pile and their reasons for placing it there. Does everyone agree? Why not?

Grades 5-12

In small groups discuss and answer each question on the *Permits* Student Page based on information gathered in previous activities. Compare answers with other groups. If there are differences, determine why.

Will a permit be needed for the planned wetland? If the answer is not clear, contact the local planning and zoning office or the state office handling wetland permits. Invite them to visit school and speak in person, or ask their permission to use a speaker phone in the classroom. Write out questions before calling and have all pertinent information about the planned wetland handy for easy reference.

Wrap-Up

Is a permit needed for the planned wetland? If so, what steps must be taken to obtain it? Is a local building permit needed?

ASSESSMENT

Were student decisions realistic? Were valid reasons given for their decisions?

EXTENSIONS

If needed, apply for the appropriate permit.

RESOURCES

Burke, D.G., E.J. Meyers, R.W. Tiner, Jr., H. Groman. 1988. *Protecting Nontidal Wetlands.* Planning Advisory Service Report Number 412/413. American Planning Association, Chicago, IL.

Environmental Law Institute. 1997. *Protecting Wetlands II: Tools for Local Governments in the Chesapeake Bay Region.* Chesapeake Bay Program, Annapolis, MD.

Environmental Law Institute. 1998. *Protecting Wetlands II: Technical and Financial Assistance Programs for Local Governments in the Chesapeake Bay Region.* Chesapeake Bay Program, Annapolis, MD.

Kusler, J. and T. Opheim. 1996. *Our National Wetland Heritage: A Protection Guide.* Environmental Law Institute, Washington, DC.

Salvesen, D. 1991. *Wetlands: Mitigating and Regulating Development Impacts.* ULI–the Urban Land Institute, Washington, DC.

On-line resources:

Army Corps of Engineers: www.usace.mil/usace-hq.html

U.S. Environmental Protection Agency:

- *Wetland Fact Sheets.* 1995. EPA843-f-95-001
- Wetland Information Hotline: 800-832-7828 or www.epa.gov
- Wetlands Division: www.epa.gov/owow/wetlands
- Surf Your Watershed: www.epa.gov/surf

Skunk cabbage, *Symplocarpus foetidus*

RULES

A group of students are on a wetland field trip. Which of these actions would make the wetland happy and which ones would make the wetland sad? Why?

Happy face	Sad face

1. Ten students walk along a wetland trail.
2. Eight students throw five stones each from the bank into the water.
3. Each student collects ten wetland plant leaves for a class project.
4. Five students quietly watch a turtle sit on a log.
5. Seven students pick up litter along the trail.
6. Four students catch butterflies and let them go.
7. Three students build a small dam of sticks and mud.
8. One student pulls nine plants out of the ground.

PERMITS

	Yes	No
1. Do NWI maps indicate that the planned wetland area is a wetland? If the answer is yes, then the wetland is a "jurisdictional wetland" subject to regulation under the Clean Water Act.		
2. Is surface water or evidence of surface water present at the planned wetland site during the growing season? If the answer is yes, the site is probably a wetland subject to regulation.		
3. Are hydric soils present at the planned wetland site? If hydric soils are present and there is evidence of water, then the site is very likely a wetland subject to regulation. If the site has hydric soils, but no signs of water, then it may have been a wetland in the past. If so, consider ways to restore the wetland hydrology, but be aware that a permit may be necessary.		
4. Are hydrophytic plants present at the planned wetland site? If obligate wetland plants are present, then the site is very likely a wetland subject to regulation. If facultative wetland plants are present, then it may be a wetland site subject to regulation.		
5. Is the wetland project larger than 1/2 acre in size? (1/2 acre = 21,780 square feet)		
6. Will vegetation be cleared before planting?		
7. Will digging, ditching, or grading be done within the wetland?		
8. Will structures such as boardwalks or overlooks be placed in the wetland?		
9. Will erosion control devices such as biologs or riprap be placed along a shoreline?		

An answer of "yes" to **any** of the first four questions above indicates the planned wetland site may be subject to wetland regulations. If this is the case and a "yes" answer was given to any of the remaining questions (5-9), then a permit will probably be needed. Locate the website or telephone number of the state agency that handles wetland permitting in your state. Contact them as soon as possible, since permitting takes several months.

If the first four questions are answered "no," then no wetland permitting is necessary. Do check with the local planning or zoning office, however, to determine if a building permit is required before any wetland construction begins.

Finding Living Benchmarks

SUMMARY

The elevation of plants relative to the water level is indicative of tide heights and seasonal water levels, because plants grow in the hydrologic zone to which they are adapted. Therefore, plants can be used as biological benchmarks of wetland hydrology. If hydrologic zones for the planned wetland are determined using biological benchmarks at nearby wetlands, an extended monitoring program of water levels is not necessary to obtain the hydrologic information necessary for wetland plantings.

OBJECTIVES

If the planned wetland has an unlimited or abundant water supply (as in tidal marshes, large lakes, river banks), then biological benchmarks will be identified and zone elevations measured. The biological benchmark data will be applied to the planned wetland, to ensure that wetland plants will be planted at appropriate elevations.

MATERIALS

- Meter sticks or folding carpenter's rule
- Line levels
- Balls of heavy string
- Brightly painted stakes (3 per team, if tidal; 4 per team, if nontidal)
- A hammer to drive the stakes
- Clipboards, paper, and pencils
- Field guides for wetland plants
- Copies of *Biological Benchmarks* Student Page (tidal or nontidal)

MAKING CONNECTIONS

Wetlands in one region of the country differ from wetlands in other regions, especially in species of plants present. To decide which plants to plant in a wetland, study similar healthy wetlands nearby. Wetland plants grow in specific zones at specific elevations relative to the water level of the wetland. Some plants tolerate standing in water, some prefer periodic flooding, others need wet soil, and still others require well-drained soil. Elevations of naturally growing native plants (relative to surface or ground water) provide biological benchmark data that can be used in planting additional plants.

BACKGROUND

Read Chapter 2, especially the section on *Unlimited or Abundant Water Supply.*

A wetland biological benchmark is provided by one or more plant species that for decades has been connected to the same water source that will be used at the planned wetland site. The elevation range occupied by benchmark plants should support the same plant community at the planned wetland site, once the site is graded and connected to the benchmark plants' water source.

The use of biological benchmarks to establish planting zones is critical for tidal emergent wetlands, tidal or nontidal submerged or floating wetlands, nontidal flood plains and nontidal wetlands associated with rivers, perennial and intermittent streams, and lakes. Biological benchmarks are generally not available for nontidal wetlands driven by groundwater or stormwater.

Biological benchmarks reflect seasonal water level changes (in lakes, rivers, and streams), tidal information (normal tidal range, spring tidal range, mean high water), and seasonal water salinities. Consequently, the hydrological information does not have to be obtained through an extended monitoring program of water levels if biological benchmarks are present at or near the planned wetland site.

PROCEDURE

Warm-Up

Can a cactus plant grow in a wetland? Why? Can a water lily grow in the desert? Why? (No. No. Plant species are adapted to specific hydrologic regimes, i.e. particular quantities of water at certain times of the year.) When plants are found growing in an undisturbed area, can they provide us information about the water available to plants in that area? (Accept all answers, and reasons. Best answer: the environment has selected which plants are better able to survive and leave offspring adapted to that niche.)

Activity

Grades K-4

Students may partner with a class of older students for the following activity, which is not suitable for most young students to do unaided. Younger students may assist others, however, by holding string and putting in the stakes.

Grades 5-12

1. To locate a reference wetland:

Find a similar healthy-looking wetland within two miles of the planned wetland site and on the same waterway, whether tidal or nontidal. Using a field guide for wetland plants and Chapter 7, identify plants representative of each zone of elevation and record them on the *Biological Benchmarks* Student Page (for tidal or nontidal areas). Check the scientific names for plants, since common names vary from place to place.

If the wetland is tidal, locate high marsh plants such as saltmeadow hay (*Spartina patens*) and groundsel tree (*Baccharis halimifolia*) on the Atlantic coast, or west coast cordgrass (*Spartina foliosa*) along the Pacific coast. Locate low marsh plants such as cordgrass (*Spartina alterniflora*) on the Atlantic coast, or Pacific sedge (*Carex opunta*) on the Pacific coast.

If the wetland is nontidal, identify plants in the aquatic zone (where plants are submerged or floating), the emergent zone (where plants typically have their feet wet), the scrub-shrub zone (woody vegetation less than 20 feet tall), and forested zone (woody vegetation more than 20 feet tall).

2. To establish transects at the reference wetland:

Working in three or more teams, have each team choose a location for a transect line that will run perpendicular to the waters' edge. Along this line the elevations at which benchmark plants are growing will be measured.

Along each transect in tidal wetlands, use a stake to mark the location closest to the water where high marsh vegetation grows. This should also be as far from the water as the low marsh vegetation grows and is called mean high water (MHW). This is the reference point that will be used to measure the elevations of all tidal plant zones. For this exercise, the elevation of the reference point will be designated as zero. Everything above this reference point is a positive elevation; everything below it is a negative elevation.

In tidal wetlands, the lowest place that low marsh vegetation grows is called mean low water (MLW). This marks the elevation of the average low tide and is where the second stake is placed. Moving up the slope, find the highest place where the high marsh vegetation is growing and mark that spot with a third stake.

In a nontidal wetland, place one stake at the water's edge. This is the reference point that will be used to measure the elevations of all plant zones. For this exercise, the elevation of the reference point will be designated as zero. Everything above this reference point is a positive elevation and everything below it is a negative elevation.

In nontidal wetlands use one stake to mark the lowest elevation for aquatic plants and another stake to mark the lowest elevation for emergent plants. Next mark the highest elevation of scrub-shrub plants with a stake; often a forested zone is above the scrub-shrub zone.

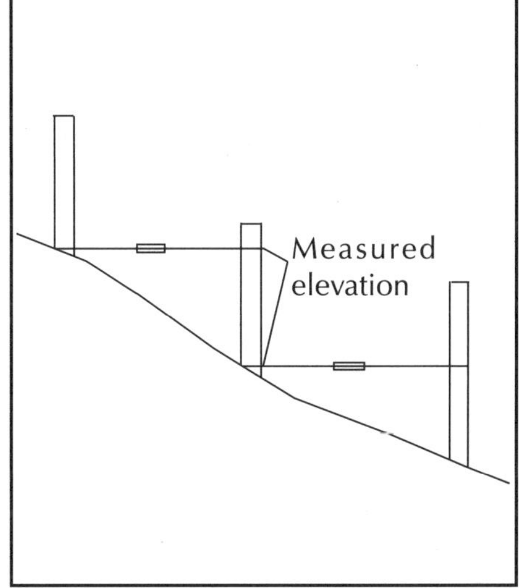

Figure 8.4 To measure elevations loop string around the base of the uphill stake, place a line level on the string, and extend the string to the meter stick placed against the downhill stake. Adjust the string at the downhill stake until the line level shows it is level. Record the elevation as measured on the meter stick.

3. To determine benchmark elevations at the reference wetland:

To measure relative elevations, begin with the stake designated as zero elevation (MHW in tidal wetlands, water's edge in nontidal wetlands). Place a meter stick or carpenter's rule next to the stake. (Make sure the low numbers are down!) Loop the end of the string around the base of the uphill stake. Extend the string to the meter stick and wrap it around once. Place a line level on the string near the middle. Adjust the level of the string at the meter stick until the line is level. Record the elevation on the appropriate *Biological Benchmarks* Student Page. Remove the string. (If elevations are determined in centimeters or inches, they must be converted to tenths of a foot for use with reference tidal data.)

Move the meter stick to the next lowest stake and loop the string end around the base of the stake designated as zero elevation. Again, extend the string to the meter stick and wrap it around once. Place the line level on the string and adjust the string at the meter stick until the line is level. Record the elevation as a negative number on the appropriate *Biological Benchmarks* Student Page. Remove the string and measure elevation at each remaining stake.

Make sketches or take pictures to help remember what the area is like. After the data has been gathered, remove all stakes and save them.

4. To calculate tide heights at the planned wetland site:

If you are dealing with a tidal area, determine the daily tidal predictions for the location closest to your planned wetland site. These may be published in a local newspaper.

a. On the internet, log onto www.co-ops.nos.noaa.gov/tp4days.html. Choose your state. Choose the location closest to your planned wetland site. Record the **Height difference for high tide** and the **Time difference for high tide**; this is the measured variation from the **Reference station.**

b. Click on the **Reference station** for your site to obtain six months of tidal predictions. Print these out for reference.

c. Go to www.tides.com on the internet. Click on **Daily Predictions**. Choose the location closest to your site for next-day tides. Record the **Mean tide range** for average tides.

d. Divide the **Mean tide range** by the **Height difference**. Round the resulting **Target high tide height** to tenths of a foot. This result is the high tide height that is needed for marking the tide zones, since the height of the high tide equals the mean tidal range.

e. Search the high tide predictions for the Reference Station. List those dates with a high tide height that matches the **Target high tide height.** Also list the times of high tide. Correct the times of high tide by adding or subtracting the **Time difference** as indicated. You will need to go to the site on one of these days to mark the tide zones, provided strong winds or storms are not present to alter the tide. Plan to go on the earliest date; so that if bad weather occurs, you will have other choices available.

If you are dealing with a nontidal flood-prone area or one with seasonal high and low water levels, determine the seasonality of high and low water levels by checking local meteorology data. Possible sources are local newspapers and weather services.

5. To establish benchmark elevations at the planned wetland site:

Go to the planned wetland site. If the site is tidal, arrive at least 30 minutes before the scheduled high tide. Sit and wait for the tide to reach its peak, then place stakes along the high tide line. This is the MHW line. By reversing the techniques used to collect the data, measure and mark the elevations for each plant zone relative to the MHW line (if tidal) or the normal water level (if nontidal) using data from the *Biological Benchmarks* Student Page. The areas marked in the planned wetland should be planted with the same types of plants as those in the reference wetland. Keep in mind that if the slope of the land surrounding the planned wetland is very different from the slope at the reference wetland, the elevations may occur farther from or closer to the water line. The distance is not important; the elevation is.

TIDE HEIGHT WORKSHEET

Reference station: ____________________

Location closest to site: ____________________

Height difference for high tide: ____________________

Time difference for high tide: ____________________

Mean tide range: ____________________

Target high tide height: ____________________

Dates high tide height at reference station matches target high tide height:

Times of daylight high tide at reference station:

Times of daylight high tide at site:

Wrap-Up

Sketch the hydrologic zones of the planned wetland site based on the biological benchmark data collected and transferred. Save all of your notes and data. Use the sketch as a guide when designing the planting area, ordering plants from a nursery, and when installing plants on the planting day. A water budget does not need to be calculated when biological benchmarks are used.

ASSESSMENT

Have students do the following:

- Match the names and pictures of some wetland plants they have encountered with the appropriate hydrologic zones.
- Describe how they measured the elevation of each zone.
- Explain why the measuring device must be held straight up (vertical).
- Explain what errors would be caused if the string is not held level.

EXTENSIONS

Investigate surveying techniques and compare them with the method used to determine biological benchmarks. How are they the same? How are they different?

RESOURCES

Cowardin, L.M., V. Carter, F.C. Golet, and E.T. LaRoe. 1979. *Classification of Wetlands and Deepwater Habitats of the United States.* U.S. Fish and Wildlife Service, Washington, DC. FWS/OBS-79/31.

Garbisch, Edgar. 1994. *Wetland Mitigation,* Section 2.5.2. Environmental Concern Inc., St. Michaels, MD.

Kesselheim, A.S. and B.E. Slattery. 1995. *WOW!: The Wonders of Wetlands.* Environmental Concern Inc., St. Michaels, MD.

Lewis, W. M. Jr., Chair. 1995. *Wetlands: Characteristics and Boundaries.* National Academy Press, Washington, DC.

On-line resources:

NOAA tide predictions at www.co-ops.nos.noaa.gov/tp4days.html

Current local tide information at www.tides.com or www.saltwatertides.com

Swamp milkweed, *Asclepias incarnata*

BIOLOGICAL BENCHMARKS--TIDAL

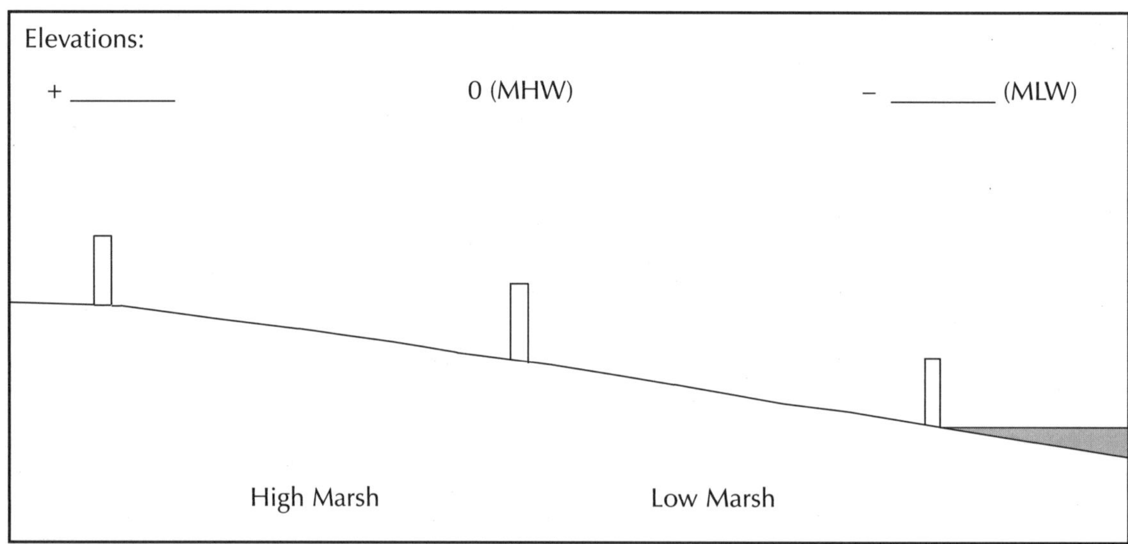

If elevations were determined in centimeters, multiply by 0.033 ft/cm, then round to the nearest tenth of a foot. If elevations were determined in inches, multiply by 0.083 ft/in, then round to the nearest tenth of a foot for use with reference tidal data.

Zone	Representative Plants
High Marsh	
Low Marsh	

BIOLOGICAL BENCHMARKS--NONTIDAL

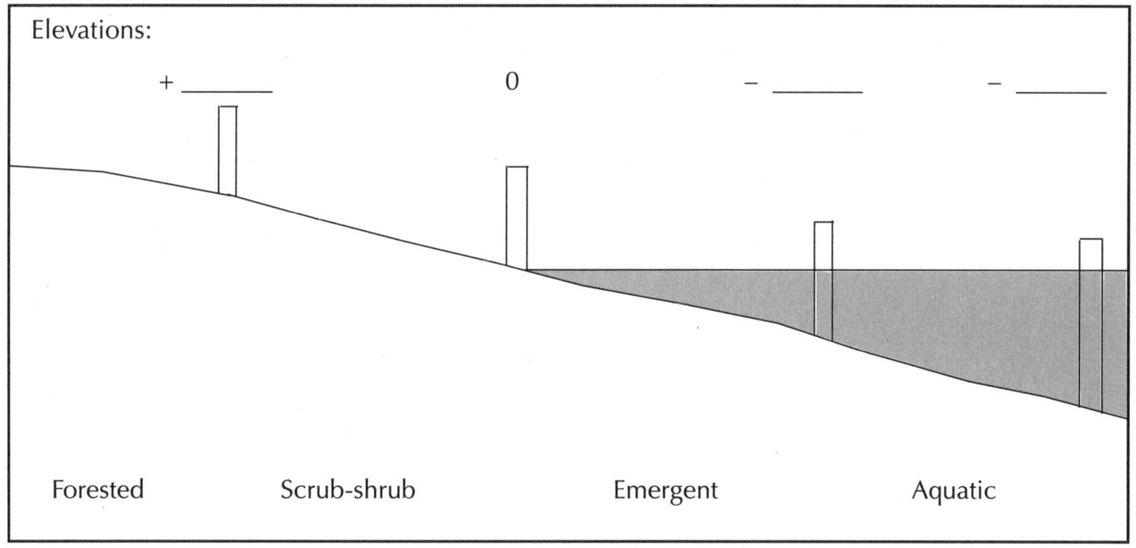

Zone	Representative Plants
Forested	
Scrub-shrub	
Emergent	
Aquatic	

Catch a Flowing Stream

SUMMARY

If a stream will provide water for the planned wetland, then stream flows must be measured to determine how much water will be available. The size of the wetland will be limited by the water available.

OBJECTIVES

If water above base flow in a perennial stream will be used for a planned wetland, then stream height above normal base flow, stream velocity, the time period of peak flow, and the width of the stream where it connects to the planned wetland will be measured. The amount of water available for the planned wetland will be calculated.

MATERIALS

- A rain gauge
- A tall stake marked in 0.1 foot increments (the stake must be higher than peak stream height)
- Two stakes for placement at each end of a relatively straight stretch of stream
- A stopwatch
- Sticks or other floating objects
- A long tape measure
- Copies of *Rainy Days* Student Page
- Copies of *Storm Data* Student Page
- Copies of *Stream Calculations* Student Pages
- Calculator (optional)

MAKING CONNECTIONS

The techniques used in this activity to measure stream flow above base levels can also be used to approximate base stream flow and the potential for flooding beyond the stream banks during storm events.

BACKGROUND

Read the section in Chapter 2 on *Unlimited or Abundant Water Supply.*

Before designing a planned wetland using stream flow, check to make sure that this is a permitted activity in your region. Some states, particularly in the West, have water rights regulations which may prevent diversion of water, even above base flow levels.

In general it is illegal to divert a stream for any reason, since it may impact resources downstream. Consequently, drawing water from streams must be done during high flow periods (storms). To determine at what height above base flow the stream will provide sufficient water to support a wetland, at least one storm event must be measured.

CAUTION: Storms can be dangerous! They are sometimes accompanied by lightning and unexpected high water. Use discretion when sending students out in a storm.

PROCEDURE

Warm-Up

Show students a measured volume of water, such as an aquarium, gallon jug, etc. Ask students how much water they think a wetland needs in order to be a wetland. (Accept all answers.) Ask students how to measure volume (length x width x height, or displacement), then ask how to measure the volume of water in a stream (discuss all ideas).

Activity

Grades K-4

What is rain? Where does it come from? Where does it go? A cross section of a stream is depicted on the *Rainy Days* Student Page. Color the pictures, cut out the squares, and then arrange the pictures in order. What word do the letters below the pictures spell? Write a sentence about what happens to a stream when it rains.

Students may assist with preparations for the following activity before the storm, and might (with an adult) do the measurements during and after the storm. Completing the *Stream Calculations* Student Pages, however, may not be appropriate.

Grades 5-12

To determine the volume of stream water available for a planned wetland, the volume of water above the stream's base flow must be assessed. The data to be measured includes:

- the width of the stream where it connects to the planned wetland,
- the amount of rainfall in a storm event,
- the peak height of the stream above base flow after a storm event,
- the stream velocity at peak height, and
- the time period during which peak flow occurs.

BEFORE THE STORM:

1. Use a tape measure to measure the width of the stream in feet where it will connect with the wetland. Record this on the *Storm Data* Student Page.

2. Install a rain gauge near the stream, but in an open area.

3. Place a stake marked in 0.1 foot intervals firmly in the ground at the edge of the water in the stream during base (normal) flow. The zero marking should be at the surface level of the stream. This stake will be used to measure the peak height of the stream during a storm.

4. Firmly place one stake at each end of a relatively straight stretch of stream. Measure the distance between these stakes in feet using the tape measure. These stakes are needed to measure the stream velocity during a storm. Record the distance on the *Storm Data* Student Page.

DURING AND AFTER THE STORM:

5. Record the time that the storm began and the time that it stopped on the *Storm Data* Student Page. Record the amount of rain collected in the rain gauge.

6. Two or three appropriately dressed students should check the marked stake at the stream edge and record the stream height hourly during and after the storm (more frequently if the stream height is changing rapidly) until the stream returns to base flow.

7. Each time the stream height data is collected, these same students should also measure and record the stream velocity. This can be done by placing a floating object (stick) in the water at the upstream stake, then timing how long it takes for the object to reach the downstream stake. Record the time in seconds on the *Storm Data* Student Page. Continue to measure stream height and stream velocity each hour until stream flow has returned to normal.

AFTER STREAM FLOW HAS RETURNED TO NORMAL:

8. Post the *Storm Data* information on the board or on an overhead transparency. Complete *Stream Calculations* to determine the volume of stream water above base flow during the storm.

Wrap-Up

Was the amount of rainfall typical for a two-week rain event?

How can the average amount of rainfall in a two-week period be determined?

The total volume of stream water above base flow that was calculated can now be entered into the water budget calculations to determine if there is enough water available naturally to support a wetland of the size planned. (See page 169.)

ASSESSMENT

Check student pages of *Stream Calculations*.

- Were their calculations accurate?
- Were correct units included for each calculation?
- Was the final answer correct?

EXTENSIONS

Alter the data and have students do additional calculations. For example: if the stream was twice as wide, what would the total water volume above base flow have been? If the time for the floating object to travel between stakes was half what was measured, what would the total water volume above base flow have been?

RESOURCES

Gilmer, M. 1995. *Living on Flood Plains and Wetlands: A Homeowner's High-Water Handbook.* Taylor Publishing Company, Dallas, TX.

The Indoor River Book. 1997. A Common Roots Guidebook. Kendall/Hunt Publishing Company, Dubuque, IA.

Tiner, R.W., and D.G. Burke. 1995. *Wetlands of Maryland.* U.S. Fish and Wildlife Service, Ecological Services, Region 5, Hadley, MA and Maryland Department of Natural Resources, Annapolis, MD.

RAINY DAYS

Place pictures of the stream in the order that they would occur.

1. Before the rain
2. As the rain begins
3. When it has rained awhile
4. After the rain stops

The letters under the pictures spell a word. What is the word? ______________

Write a sentence about what happens to a stream when it rains.

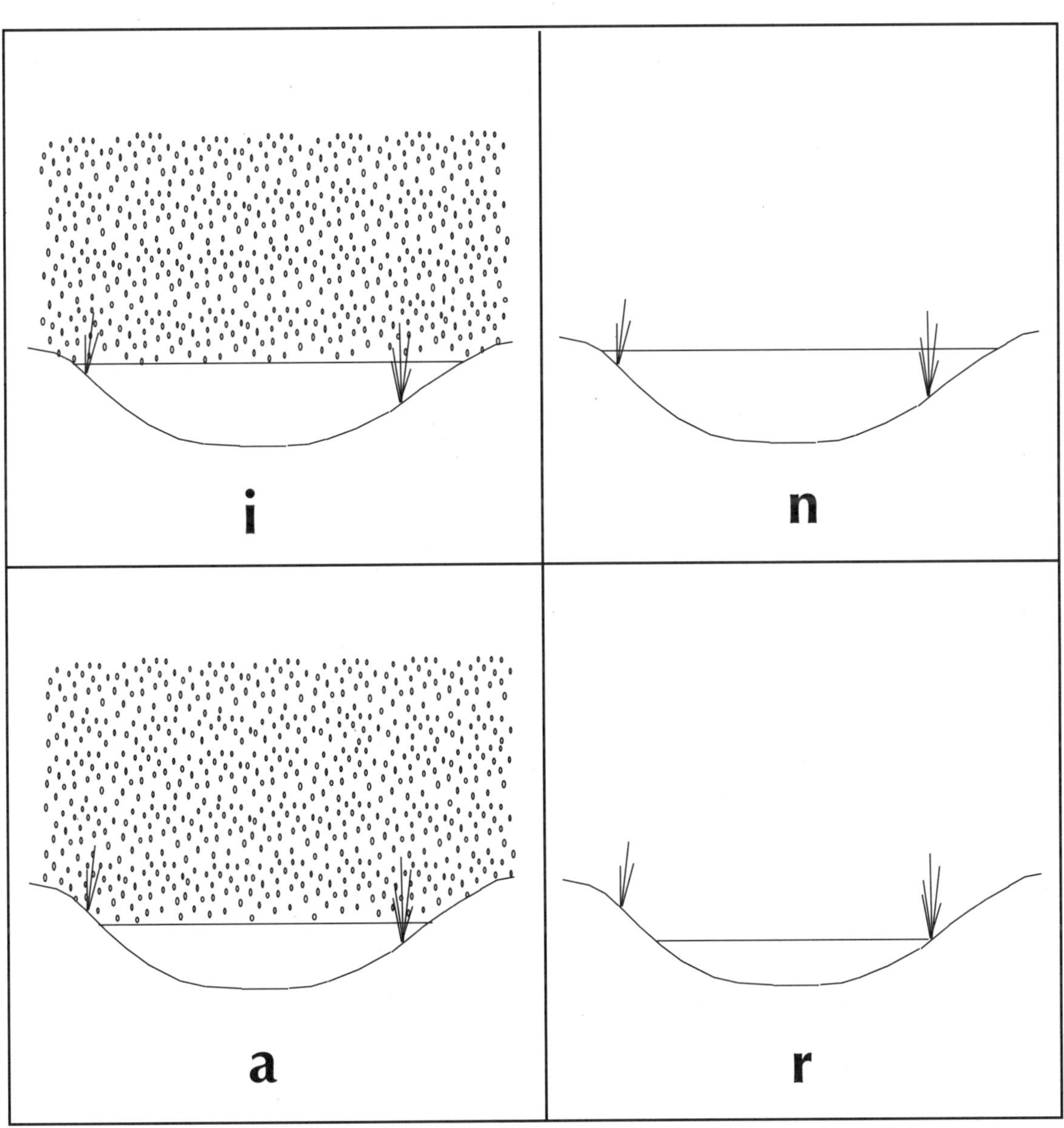

STORM DATA

BEFORE THE STORM:

1. The stream width is __________ feet.

2. The distance between the two stakes is _____________ feet.

DURING THE STORM:

3. The storm began at ____________ AM/PM on _______________.

4a. One hour later at ______AM/PM the height of the water was _______ feet, and the float took _________ seconds to travel between the stakes.

4b. One hour later at ______AM/PM the height of the water was _______ feet, and the float took _________ seconds to travel between the stakes.

4c. One hour later at ______AM/PM the height of the water was _______ feet, and the float took _________ seconds to travel between the stakes.

4d. One hour later at ______AM/PM the height of the water was _______ feet, and the float took _________ seconds to travel between the stakes.

4e. One hour later at ______AM/PM the height of the water was _______ feet, and the float took _________ seconds to travel between the stakes.

4f. One hour later at ______AM/PM the height of the water was _______ feet, and the float took _________ seconds to travel between the stakes.

4g. One hour later at ______AM/PM the height of the water was _______ feet, and the float took _________ seconds to travel between the stakes.

Continue on the back of this paper if necessary.

5. The storm ended at ____________ AM/PM.

6. The rain gauge contained ____________ inches of water.

STREAM CALCULATIONS

Use the information from the *Storm Data* Student Page to make these calculations. #1, #2, #3, #4, #5, and #6 correspond to the numbers on the data sheet. Don't forget your units!

A. How long did it rain? Subtract the time started from the time ended. #5 – #3 = ?

B. To determine the rate of precipitation divide the amount of rain (in inches) in the rain gauge by the number of hours of rain. #6 ÷ #A = ?

C. Calculate the stream velocity each hour by dividing the distance between stakes (in feet) by the time (in seconds) that it took the float to travel between them. #2 ÷ #4 = ?

a.

b.

c.

d.

e.

f.

g.

D. Calculate the water above base flow each hour by multiplying the stream width (in feet) times the height of the water (in feet) times the stream velocity (in feet per second) times the time (in seconds). #1 x #4 x #C x 3,600 = ? (How many seconds are there in one hour?)

a.

b.

c.

d.

e.

f.

g.

E. Now calculate the total water volume above base flow by adding together all of your answers in #D. $\#D_a + \#D_b + \#D_c + \#D_d + \#D_e + \#D_f + \#D_g = ?$

F. How long did it take to reach maximum stream height?

G. How long did it take for the stream to return to its base flow level?

H. Is this enough water to fill the planned wetland to its pool level (the planned water depth)?

Will Stormwater Run Off?

SUMMARY

The stormwater runoff that is available to support a planned wetland will be calculated. The quantity of water runoff will then be available for calculating a wetland water budget.

OBJECTIVES

If stormwater runoff is available for a planned wetland, then students will (a) locate outfalls from storm drains, (b) define the drainage area, and (c) calculate stormwater runoff from the watershed to the planned wetland.

MATERIALS

- Topographic map of the watershed
- Precipitation data during the local growing season
- School building site plan
- Graph paper (transparent)
- Food coloring
- **Table 3** and **Table 4** (pages 41 and 42)
- Copies of *Stormwater Runoff* Student Pages
- Calculator (optional)

MAKING CONNECTIONS

Through hands-on activities and mathematical calculations, students will solve practical problems.

BACKGROUND

Read Chapter 2, especially the section on *Limited or Uncertain Water Supply* (pages 35-44)

One of the more common planned wetlands in schoolyards and neighborhoods is the "rain garden," which is entirely dependent on stormwater for its water supply. The watershed supplying the stormwater might be the portion of a school roof that directs rain water to a particular downspout, or parking lot water directed to a storm drain that could be intercepted. The watershed might be the yards and streets of a housing development, or the school playing fields that direct water flow down a grassy swale. These small local watersheds might supply enough water to support a planned wetland.

PROCEDURE

Warm-Up

Where does the water go when it rains on the school roof? Where does rainwater go when it rains on the parking lot or the playing fields? Take a walk outside to investigate. Try to follow the path of the rainwater from where it lands until it meets up with another body of water. To test where water goes when it flows down a drain, pour a pint jar of water colored with an entire bottle of food coloring down a drain (additional water will be needed if the drain is dry). Station students at various possible outfalls to watch for the now-diluted colored water to appear. (Be aware that some students are color blind and the color change will not be obvious to them.) Test all the drains that you might use to supply your planned wetland.

Activity

Grades K-4

Repeat the warm-up activity to allow all students to see where the colored water goes. If students have not mastered the math skills needed, older students or teachers can either assist or do the actual calculations of stormwater runoff.

Grades 5-12

Define the drainage area for the planned wetland as directed on page 36. Using a topo map, start at the discharge point of the planned wetland. Draw the boundary by starting on one side of the discharge point. Move uphill from contour line to contour line by the shortest possible path. Cross each contour line at a right angle. The line that you are drawing is a ridge line. When you reach the topographic high point, stop working on that line and start again at the discharge point going uphill on the other side of the discharge point. If you connect at the high point with your first line, the drainage area boundary is complete. If your lines do not connect, then you have reached two separate topographic high knobs, which then must be connected through what is called a saddle point.

To calculate the total drainage area and then the area of each Land Use/ Surface Cover found in the drainage area, overlay a thin sheet of graph paper or a transparency of the graph paper provided on page 166. Trace the entire drainage area on the graph paper (or transparency), then count the number of squares within that area. Place the graph paper against the scale of the map to determine the distance and area represented by each square. Repeat with each Land Use area.

To determine the average monthly precipitation during the growing season, check with the local Soil Conservation Service office or the local Natural Resources Conservation Service office. The same information is available on the internet (see Resources). On the table of temperature and precipitation data, mark the months where the average daily minimum temperature is above 32°F. This is the growing season. Add the average monthly rainfall for those months and divide by the number of months. This is the average monthly precipitation for the growing season.

Follow the directions on the worksheet to calculate the amount of stormwater runoff that will be available for a planned wetland on a biweekly basis.

Wrap-Up

Is sufficient water available for the planned wetland? A brief discussion of reasons for differing opinions may be appropriate. What additional information is needed?

The stormwater runoff that was calculated can now be entered into the water budget calculations to determine if there is enough water available naturally to support a wetland of the size planned. (See page 169.)

ASSESSMENT

Check student pages of *Stormwater Runoff Calculations*.

- Were student calculations accurate?
- Was their final answer correct?

EXTENSIONS

Solve the following problems:

(A) If only half the normal rainfall occurred during the growing season, would the wetland survive?

(B) If twice the normal amount of rain fell during the growing season, where would the excess water go? Would this cause problems?

RESOURCES

Aquatic Project WILD. 1992. Project WILD, Bethesda, MD.

Engineering Field Handbook; Chapter 13: Wetland Restoration, Enhancement, or Creation. 1992. USDA Soil Conservation Service. 210-EFH, 1/92.

Give Water A Hand Action Guide. 1996. The Blue Thumb Program, University of Wisconsin, Madison, WI.

On-line resources:

Local monthly precipitation data: www.wcc.nrcs.usda.gov/water/wetlands.html

Mallard drake,
Anas platyrhynchos

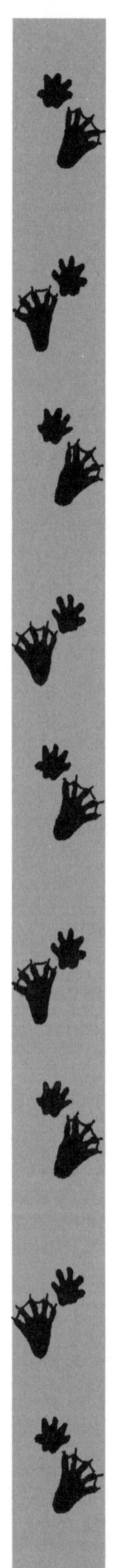

STORMWATER RUNOFF CALCULATIONS

Directions	Calculations	Answers
1. Calculate the area (in acres) of the watershed using the topographic map and/or the school site plan. (1 acre = 43,560 square feet)		a = acres
2. Calculate the area of each Land Use/Surface Cover found in the drainage area, including the planned wetland using the topo map and/or site plan for the school. See **Table 4** for a list of land uses.		a_s = a_t = a_u = a_v = a_w = square feet
3. Convert the area of each Land Use/Surface Cover to a percentage of the whole area. Use decimals (0.12, not 12%).		a_s = a_t = a_u = a_v = a_w = percentages
4. Determine the Hydrologic Soil Group for each Land Use area by sampling each area as in *What Grows There?* and comparing with **Table 3**.		s = t = u = v = w =

Directions	Calculations	Answers
5. Using **Table 4**, determine the Runoff Coefficient for each Land Use area.		c_s = c_t = c_u = c_v = c_w =
6. Multiply the Runoff Coefficient for each Land Use area by the percentage of the watershed covered by that Land Use. Add all of these together for the Runoff Coefficient. $c = c_s a_s + c_t a_t + \ldots = ?$		c = not adjusted
7. Determine the average monthly precipitation during the local growing season; then divide by 2.		i = inches
8. If i $\leq$ 2 inches, reduce c (#6) by 35%. if i >2 inches, but $\leq$ 3 inches, reduce c by 20%.		c = adjusted
9. Calculate runoff in the drainage area containing the planned wetland by multiplying the drainage area in acres (#1) times the runoff coefficient (#8) times the average two-week precipitation during the growing season (#7) times 3,630 cubic feet/acre inch. R_{in} = aci(3630 cu ft/acre in) = ?		R_{in} = cubic feet biweekly

Balancing Water Budgets

SUMMARY

The water budget for a planned wetland takes into account all water coming into (input) and all water going out of (output) the wetland. It includes surface water and groundwater, and is calculated based on data collected in earlier activities and on measurements determined by others through research.

OBJECTIVES

Students will calculate the water budget for a planned wetland and determine whether the water supply is sufficient for the size wetland planned.

MATERIALS

- Data from *Stream Calculations* Student Pages, if applicable
- Data from *Stormwater Runoff Calculations* Student Pages, if applicable
- Data on the size of the planned wetland (length, width, average depth)
- Map or globe showing latitude
- Temperature data for the local growing season
- Copies of *Water Budget* Student Pages
- **Table 1** and **Table 2** (pages 39 and 40)
- Calculator (optional)

MAKING CONNECTIONS

Like professionals, students determine if there is sufficient water available for their planned wetland. In similar ways, urban planners determine whether drinking water supplies are adequate and when water rationing is needed. Farmers also make decisions about irrigation of crop land using similar calculations.

BACKGROUND

Read Chapter 2, especially the section on *Limited or Uncertain Water Supply* (pages 35-44).

Without sufficient water, wetlands cannot exist. Some wetlands have unlimited and reliable water supplies; a water budget need not be calculated for these. Other wetlands have a potentially limited or unreliable water supply that must be evaluated to determine whether there is sufficient water to support a planned wetland.

Tidal saltwater wetlands have an unlimited supply of ocean water. Tidal freshwater wetlands and nontidal wetlands connected to rivers, lakes, and perennial streams associated with large watersheds have abundant water supplies and a water budget need not be calculated. Also, if local water will be supplied by a hose to a small planned wetland, then a water budget need not be determined.

A water budget should be calculated for wetlands dependent on stormwater runoff, stream flow, groundwater-fed springs, direct precipitation, and any combination of these water sources. If the water supply is unpredictable, the wetland should be planned to have more water than is likely necessary to maintain it. The actual calculation of a water budget will help determine if the water supply is grossly inadequate and whether modification of wetland size or water supply will be necessary.

The most basic form for a water budget is:

Water Storage = Water Input – Water Output

The more detailed form that we will use is:

Change in Water Storage (Δ S)	=	Precipitation (P)	+	Runoff (R_{in})	+	Groundwater Discharge (W_g)
	–	Water Out (W_{out})	–	Infiltration (I)	–	Evapotranspiration (ET)

PROCEDURE

Warm-Up

Ask students to think of their piggy banks or bank accounts. If more money is being put into the bank than is being withdrawn, what happens? If more money is being withdrawn than is being put into the account, what happens? Why do we budget our money? How is this like budgeting our water supply?

Activity

Grades K-4

Younger students should participate in the Warm-Up and the calculations according to their ability levels. The requisite math computation skills for this activity are the ability to add, subtract, multiply, and divide (or use a calculator to perform these functions). For the calculations that students are unable to perform, allow them to take turns entering numbers calculated by adults or older students on an enlarged table. Explain to students the use of a letter (t) as a symbol for a word (temperature).

As in the piggy bank analogy, use a glass container to represent a wetland. Marbles, marshmallows, or other objects might represent the quantity of water going into and out of a wetland so students can visualize the calculations made. Be sure all objects represent the same amount of water.

Grades 5-12

Complete the *Water Budget* Student Pages if the planned wetland will be supported by stormwater runoff, stream flow, direct precipitation, or a groundwater spring, or if a combination of these water sources is planned.

To determine the average monthly precipitation for the area, check the local Soil Conservation Service office or the local Natural Resources Conservation Service office. The same information is available on the internet (see Resources). On the table of temperature and precipitation data, mark the months where the average daily minimum temperature is above 32°F. This is the growing season. Add the average monthly rainfall for those months and divide by the number of months. This is the average monthly precipitation for the local growing season.

To determine the average monthly temperature for the local growing season, use the same table as above. For each month during the growing season, add together the average monthly temperatures, then divide by the number of months in the growing season. This is the average daily temperature for the entire growing season.

Wrap-Up

Based on your water budget calculations, will there be enough water to fill the wetland and keep it full? If the answer is yes, then continue planning. If the answer is no, then discuss possible changes to the planned wetland that will provide sufficient water to maintain it. (Are there additional water sources? Could the size be reduced?) Work through the Sizing Exercise on page 41.

After plants have become well established, a permanently flooded wetland can tolerate dry conditions for about two weeks. A regularly flooded wetland can tolerate dry conditions for about four weeks; and an irregularly or seasonally flooded wetland can tolerate dry conditions for about eight weeks. It would be best, however, for there to be an excess of water available to the wetland so that the wetland does not fail.

ASSESSMENT

Review student pages of the *Water Budget.*

- Were calculations accurate?
- Were conclusions about whether the available water would support a wetland valid and reasonable?
- Were modifications suggested, and were they reasonable?

EXTENSIONS

If rainfall during the growing season is twice the normal average, predict what would happen to the planned wetland. If an unusual drought occurred and only half the normal average rainfall occurred, would the wetland survive? Why? What could be done to protect the wetland?

RESOURCES

Engineering Field Handbook; Chapter 13: Wetland Restoration, Enhancement, or Creation. 1992. USDA Soil Conservation Service. 210-EFH, 1/92.

Garbisch, E. W. 1994. *Achieving the Correct Hydrology to Support Constructed Wetlands.* Environmental Concern Inc., St. Michaels, MD.

Gardner, R.C. 1997. *Rainfall and Runoff.* Longwell Run Watershed Management Project, MD DNR, Annapolis, MD.

Kelmelis, J.A. 1994. *A Blueprint for Change, Part V: Science for Floodplain Management into the 21st Century.* SAST, USGS, Washington, DC.

Mitsch, W.J. and J.G. Gosselink. 1993. *Wetlands.* Von Nostrand Reinhold, New York, NY.

On-line resources:
Local monthly temperature data: www.wcc.nrcs.usda.gov/water/wetlands.html

Buttonbush,
Cephalanthus occidentalis

WATER BUDGET

Water Sources

1. If a planned wetland will be supported by direct precipitation, then determine the average monthly precipitation during the growing season. Divide by 2 for biweekly precipitation; divide by 12 to convert to feet. Multiply by the area (length times width) of the wetland for the precipitation in cubic feet biweekly.

If a planned wetland will be supported by stormwater runoff and is located within its own drainage area, then precipitation has been calculated and included in R_{in} and should not be added in a second time, therefore P = 0.

Calculations:

P =   cubic feet biweekly

2. If a planned wetland will be supported by stormwater runoff, then record the amount calculated for R_{in} on the Student Page *Stormwater Runoff Calculations.*

If a planned wetland will be supported by stream flow, then record the total water volume above base flow (#E) from the Student Page *Stream Calculations,* if this was a typical two-week period. (Was the amount of rain collected in the rain gauge the same as the biweekly precipitation calculated in #1?)

If a planned wetland will not receive any water from runoff, then R_{in} = 0.

R_{in} =   cubic feet biweekly

3. If a planned wetland will be supported by groundwater, then measure the volume of flow and calculate the biweekly volume. (Professional assistance may be required.)

If groundwater will not contribute any water to the planned wetland, then W_g = 0.

W_g =   cubic feet biweekly

Water Losses

4. Determine the volume of water necessary to fill the wetland to the designed pool level. Multiply the area of the wetland in square feet (length times width) times the average water depth.

Calculations:

pool volume =_______ cubic feet

5. The water leaving the wetland equals R_{in} (#2) minus the volume of the wetland at pool level (#4). If R_{in} is less than the volume of the wetland at pool level, then $W_{out} = 0$.

Calculations:

W_{out} =   cubic feet biweekly

6. Determine the approximate latitude for the planned wetland (use a globe or topo map). Determine the average daytime hours for the months during the growing season, given as a percentage of the year from **Table 1**.

p = ________ percentage

7. Determine the average monthly temperature in °F for the local growing season.

t = _______°F

8a. The consumptive use coefficient for rice is 1.2 inches per month. The conversion factor is 36,000 days per month inches per foot. Calculate daily water loss through evaporation and transpiration using the formula ET = 1.2 x p x t ÷ 36,000.

Calculate:

ET = _______ feet/day

8b. Multiply ET in feet/day times the area of the wetland (length x width) times 14 to determine ET in cubic feet biweekly.

Calculate:

ET =
cubic feet biweekly

9a. Check soil texture from the *Soil Transect* Student Page. Determine the rate at which water soaks into the ground (infiltration) from **Table 2**.

I = _______ feet/day

9b. Multiply I in feet/day times the area of the wetland (length x width) times 14 to determine I in cubic feet biweekly.

Calculate:

I =
cubic feet biweekly

The Water Budget

10. To determine the amount of water that will be available for the planned wetland (water storage) add all the water sources ($P + R_{in} + W_g$) and subtract all the water losses (W_{out} - ET - I) using the following formula: $\Delta S = P + R_{in} + W_g - W_{out} - ET - I$.

ΔS =
cubic feet biweekly

If insufficient water is available for the planned wetland, then additional water may be supplied by hose or the wetland size may be reduced.

9 PLANNING A WETLAND

ACTIVITIES

Discovering Dips

Making the Grade

Planting Plans

Clarifying Specs

Discovering Dips

SUMMARY

Using survey techniques, students will measure elevations for an area that will include the planned wetland. The elevation data will be recorded in a grid pattern and contour lines drawn to form a topographic map of the planned wetland area.

OBJECTIVES

Students will survey elevations and create a topographic map of the planned wetland area. Three-dimensional contours will be translated into a two-dimensional topographic map.

MATERIALS

- A level or transit with a tripod (borrow or rent)
- An elevation rod (borrow or rent)
- A measuring wheel (borrow or rent) or a 50-foot/15-meter measuring tape
- A pocket compass
- A 2" x 2" wood stake, a large nail, and colored survey tape or ribbon for each benchmark that must be created OR paint to mark an existing permanent structure
- Survey flags or wire flags
- Copies of *Rod Readings* Student Page
- Copies of *Ground Elevations* Student Page
- Copies of appropriate *Topographic Map* Student Page
- Copies of *Estimation Skills* Student Pages

MAKING CONNECTIONS

The techniques used by surveyors to measure elevations and create topographic maps will be practiced by students. These job skills are used in construction projects, such as buildings and roads. Elevations are determined before construction and after construction (these are called "as built" conditions), and are used to determine changes in elevation over time. Homeowners and landowners use survey information to determine the boundaries of their property and to place structures (buildings, wells, sewer lines, etc.) on their property.

BACKGROUND

Read Chapter 4, especially the sections about *Elementary Topographic Surveying* and *Drawing a Topographic Map* on pages 55-57.

Among their many accomplishments, George Washington and Merriweather Lewis were surveyors. While the tools have improved, the techniques are much the same. A modern surveyor sights distances and elevations using a survey instrument equipped with a laser and computer data pack. Information stored in the data pack is downloaded to a computer in the office, then analyzed with a computer program to produce contour lines and to plot a topographic map with an electronic plotter. For the small area to be surveyed, such expensive equipment is not needed, but the techniques are the same.

PROCEDURE

Warm-Up

Has anyone seen metal discs imbedded in concrete markers by the side of a street, or nails with bright colored plastic tape hammered into pavement, or wooden stakes with brightly colored plastic ribbon tied on them? These are evidence that a surveyor has been at work. Was a road or building constructed at the site?

Activity

Grades K-4

If a road or building is planned nearby, arrange for students to watch the surveyors in action. The surveyors may be willing to show their equipment to the students and possibly guide them in doing their planned wetland survey.

As an alternative activity, use clay to model the contours of the area that others will survey. Show high areas and low areas. Mark the area for the planned wetland.

The survey activity that follows can be done by younger students with older students or parents as partners. Consider teaming a class of younger students with a class of older students in a buddy system to survey the planned wetland area and create the topographic map.

Grades 5-12

Rent or borrow the needed equipment and become familiar with it. A parent, friend, or relative may have access to survey equipment. Also check with the county or municipal engineers; not only do they have survey equipment, but they may be willing to demonstrate its use and assist with the actual survey. Also consult with the local office of the Natural Resources Conservation Service for advice and possible equipment.

Most groups will be too large for everyone to participate in the survey at the same time. While teams take turns doing the survey, others can practice estimation skills. Place five markers or flags at varying distances apart, but outside the survey area. (The distance between the first two flags should be fifteen meters or fifty feet.) Using the *Estimation Skills* Student Pages, all students can determine their pace over a measured distance, then estimate distances between subsequent markers. Actual distances will be measured after everyone has walked the course.

1. With student assistance, locate an existing survey benchmark within sight of the planned wetland. A survey benchmark was probably used during construction of the school. Determine the elevation of the benchmark by reading its inscription, and/or contacting the office that installed it when the earlier survey was made. If an existing benchmark cannot be found, select a location and create a project benchmark. (See Step 1, page 55.) Assign the new benchmark an elevation (such as 50 feet) that will allow all measurements to be positive numbers.

2. Set the transit on the tripod at a location where the benchmark and all of the area to be surveyed can be seen. Level the instrument first by adjusting the tripod legs, and then by using the fine adjustment knobs just below the transit. If it becomes necessary to move the transit, review *Moving the Survey Instrument*, page 56. Assign people to measure distances for placement of survey flags (distance person), mark stations with survey flags (flag person), survey (instrument person), hold the rod (rod person), and record elevations (recorder).

3. Establish a linear baseline along one side of the area to be surveyed. If the baseline parallels a road, sidewalk, building, or other permanent structure, then replication of these measurements will be much easier. Place a survey flag at the interval selected along the baseline (10 feet is suggested), beginning at Station 0+00 and continuing across the area to be surveyed. See **Figure 9.1**.

4. Use the pocket compass to determine the bearing of the baseline by sighting along the station flags that have just been set along the baseline. Record the compass bearing on the *Rod Readings* Student Page. Stand at Station 0+00 with the compass. Turn 90° from the baseline. The flag person working with the distance person should now place flags at the selected interval (10 feet) to form a transect, perpendicular to the baseline, that continues across the area to be surveyed. Repeat at Station 0+10, Station 0+20, etc. until a grid pattern of flagged stations has been laid out. See **Figure 9.2**.

5. Now the rod person places the survey rod vertically on the project benchmark, not beside it! The instrument person then focuses the instrument on the rod and reads the elevation. Record this as the rod reading (RR) for the benchmark on the *Rod Readings* Student Page. The recorder then adds the benchmark elevation (BE) to the RR to obtain the height of instrument (HI) as in **Equation 4.1** (page 56). If the BE is not known, assign an elevation such as 50 feet.

6. The rod person stands at Station 0+00 and holds the rod vertically next to the flag. The instrument person sights on the rod and states the RR. The recorder writes the RR for the station on the *Rod Readings* Student Page. Repeat this at each flag along the perpendicular transect from Station 0+00. In the same manner measure the elevations at Station 0+10 and each flag along the perpendicular transect from that station. Repeat until RR's have been obtained for each flag as in **Figure 9.3**.

7. Using the information recorded on the *Rod Readings* Student Page, for each station and each flag subtract the rod reading (RR) from the height of the instrument (HI) to obtain the ground elevation (GE). Record these numbers on the *Ground Elevations* Student Page as in **Figure 9.4**.

8. To create a topographic (topo) map of the area, individually or in small groups transfer the ground elevation numbers from the *Ground Elevations* page onto the appropriate *Topographic Map* Student Page.

9. Draw in contour lines at 1.0 foot intervals. **Figure 9.5** shows one example in which the elevation drops along the baseline from 47.1 to 46.8. Since there is a whole number in between, a contour line for 47.0

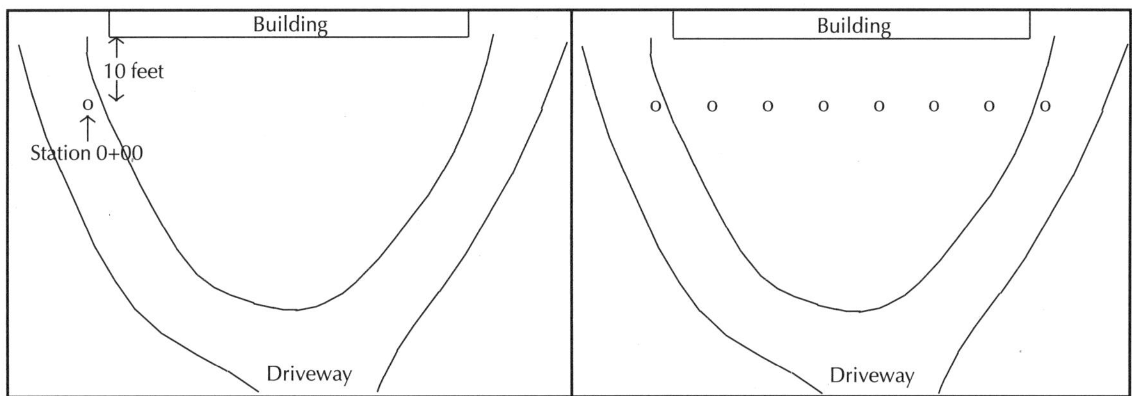

Figure 9.1 Flags were placed at Station 0+00 and along the baseline.

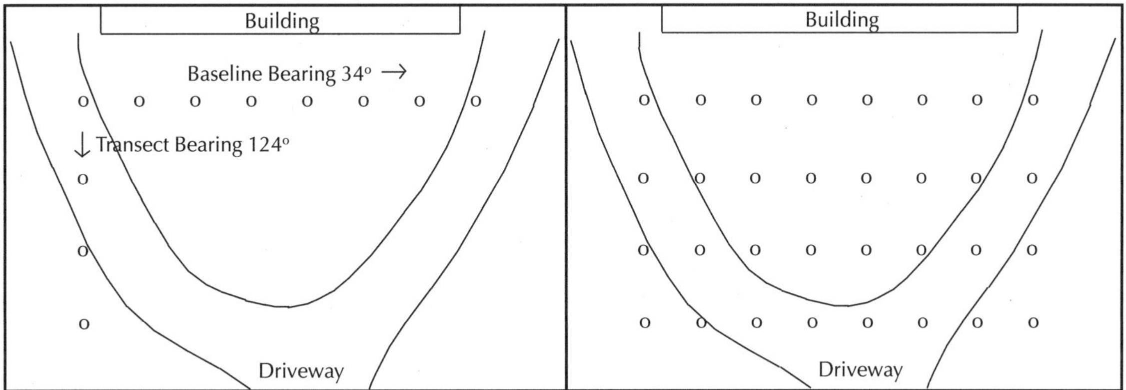

Figure 9.2 Transect flags were placed at 90° angles from the baseline.

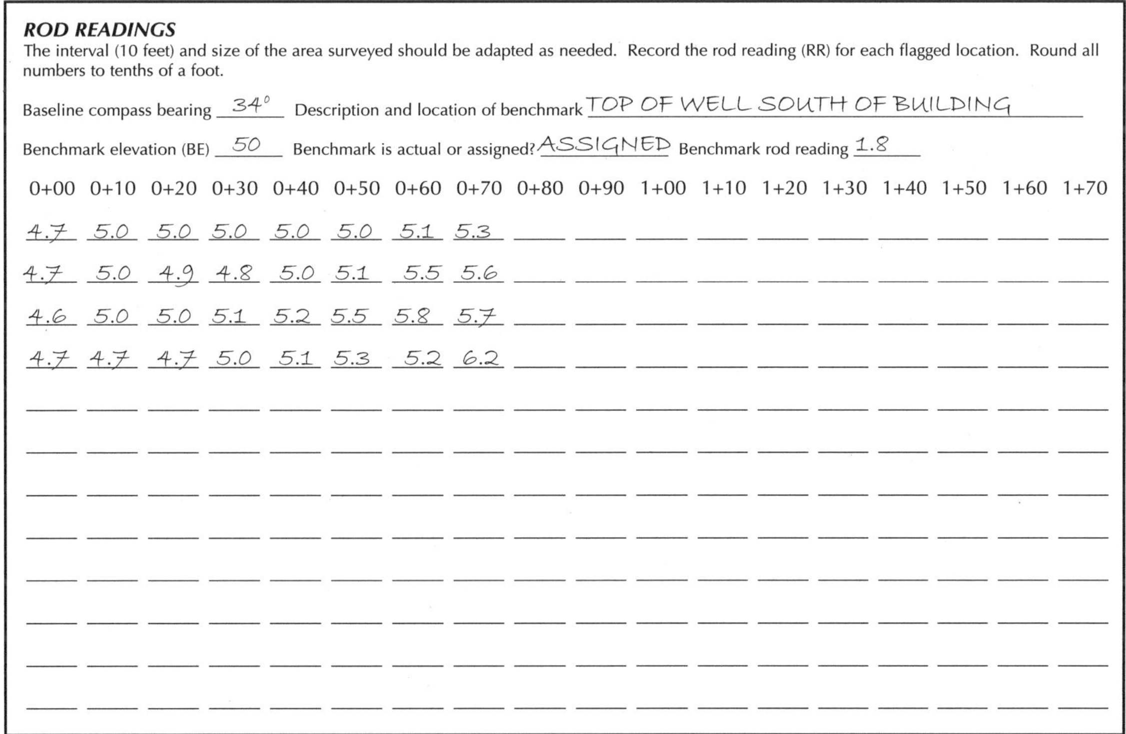

ROD READINGS

The interval (10 feet) and size of the area surveyed should be adapted as needed. Record the rod reading (RR) for each flagged location. Round all numbers to tenths of a foot.

Baseline compass bearing 34° Description and location of benchmark TOP OF WELL SOUTH OF BUILDING

Benchmark elevation (BE) 50 Benchmark is actual or assigned? ASSIGNED Benchmark rod reading 1.8

0+00	0+10	0+20	0+30	0+40	0+50	0+60	0+70	0+80	0+90	1+00	1+10	1+20	1+30	1+40	1+50	1+60	1+70
4.7	5.0	5.0	5.0	5.0	5.0	5.1	5.3										
4.7	5.0	4.9	4.8	5.0	5.1	5.5	5.6										
4.6	5.0	5.0	5.1	5.2	5.5	5.8	5.7										
4.7	4.7	4.7	5.0	5.1	5.3	5.2	6.2										

Figure 9.3 The *Rod Readings* above are from a project at Horsehead Wetlands Center.

GROUND ELEVATIONS

The interval (10 feet) and size of the area surveyed should be adapted as needed. Calculate the height of the survey instrument (HI), then calculate the ground elevation (GE) for each flagged location using the data recorded on the *Rod Readings* Student Page. Round all numbers to tenths of a foot.

HI = Benchmark RR + BE = 51.8 GE = HI – RR (for each station or flag)

0+00	0+10	0+20	0+30	0+40	0+50	0+60	0+70	0+80	0+90	1+00	1+10	1+20	1+30	1+40	1+50	1+60	1+70
47.1	46.8	46.8	48.8	46.8	46.8	46.7	46.5										
47.1	46.8	46.9	47.0	46.8	46.7	46.3	46.2										
47.2	46.8	46.8	46.7	46.6	46.3	46.0	46.1										
47.1	47.1	47.1	46.8	46.7	46.5	46.6	45.6										

Figure 9.4 The *Ground Elevations* above are from a project at Horsehead Wetlands Center.

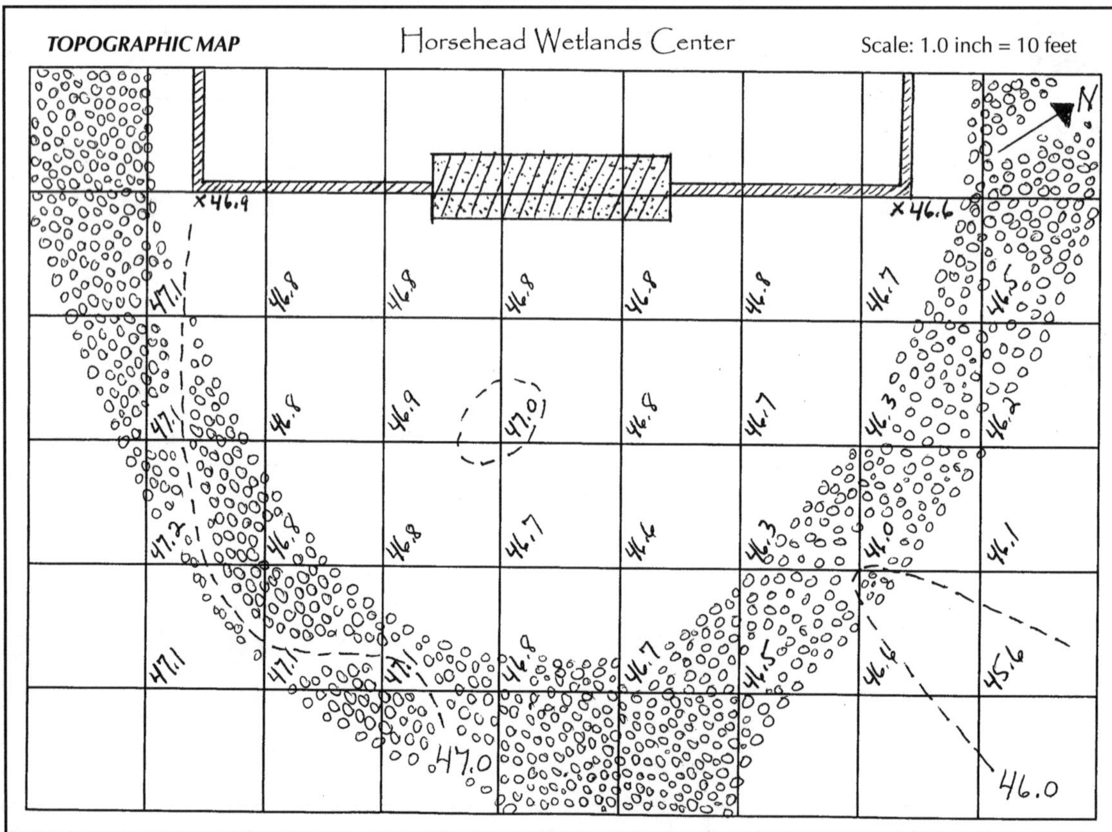

Figure 9.5 This is the *Topographic Map* for the project at Horsehead Wetlands Center.

feet is marked in pencil. In the lower right corner of the example, a contour line for 46.0 feet is marked on what is nearly flat land that gently slopes to the lower right of the topo map. One isolated high spot of 47.0 feet is marked.

Where the land has steep slopes, the contour lines will be close together. Where the land is flatter, contour lines will be farther apart. Label contour lines with the elevation, such as 10 feet, 11 feet, 12 feet, etc. Also mark the scale. One block on the *Topographic Map* Student Page equals ten feet, if that is the survey interval used. Two *Topographic Map* Student Pages are supplied. The page with the larger grid is for areas smaller than 90 feet by 60 feet; the other page is for larger areas.

Wrap-Up

The topo maps you have just created will be used to prepare a grading plan in the next activity. Do the topo maps agree with what was seen when looking at the surveyed area? Are high spots high? Are low spots low? If not, why not? (If necessary, review procedures to determine how errors might have occurred.)

ASSESSMENT

- Were students successful in laying out the area to be surveyed?
- Did the ground elevation numbers make sense when students added contour lines to their maps?
- Compare student-produced topographic maps. Are they all alike? If not, why not? A common problem is placement of contour lines on the wrong side of the mark representing the station or flag.

EXTENSIONS

After the grading plan is completed in the next activity, return to the surveyed area to mark on the ground with stakes, flags, and spray paint the changes that will be made to the land contours.

RESOURCES

USDA-SCS. 1992. "Chapter 13: Wetland Restoration, Enhancement, or Creation", *Engineering Field Handbook*. U.S. Department of Agriculture, Soil Conservation Service, Washington, DC. (210-EFH, 1/92)

Wirshing, J.R. and R. H. Wirshing. 1985. *Schaum's Outlines: Introductory Surveying*. McGraw-Hill, Washington, DC.

ROD READINGS

The interval (10 feet) and size of the area surveyed should be adapted as needed. Record the rod reading (RR) for each flagged location. Round all numbers to tenths of a foot.

Baseline compass bearing ________ Description and location of benchmark ________________________________

Benchmark elevation (BE) ________ Benchmark is actual or assigned? ______________ Benchmark rod reading ________

0+00	0+10	0+20	0+30	0+40	0+50	0+60	0+70	0+80	0+90	1+00	1+10	1+20	1+30	1+40	1+50	1+60	1+70

GROUND ELEVATIONS

The interval (10 feet) and size of the area surveyed should be adapted as needed. Calculate the height of the survey instrument (HI), then calculate the ground elevation (GE) for each flagged location using the data recorded on the *Rod Readings* Student Page. Round all numbers to tenths of a foot.

HI = Benchmark RR + BE = _______ GE = HI – RR (for each station or flag)

0+00	0+10	0+20	0+30	0+40	0+50	0+60	0+70	0+80	0+90	1+00	1+10	1+20	1+30	1+40	1+50	1+60	1+70

TOPOGRAPHIC MAP *(for areas smaller than 90' x 60')*

Scale: 1.0 inch = ______ feet

TOPOGRAPHIC MAP (for large areas)

Scale: 0.5 inch = ______ feet

ESTIMATION SKILLS

Preparation: Using a measuring tape, mark the beginning and end of a 15-meter (or 50-foot) distance with flags labeled A and B. From flag B, change direction and walk a distance, then mark the location with flag C. Do the same for flags D and E. Do this around the perimeter of the area being surveyed.

1. The distance from flag A to flag B is 15 meters. Walk from flag A to flag B using a normal walking pace, not a long one! Carefully count steps. Do this three times, then calculate the average.

   On average, I take _______ paces to go from flag A to flag B.

   My pace is ________ meters long. (15 m ÷ number of paces)

2. My guess for the distance from flag B to flag C is _________ meters.

   The number of paces counted from flag B to flag C is _________ paces.

   My estimate of the distance from flag B to flag C is __________ meters.

   (Multiply the number of paces times the length of your pace calculated in #1.)

3. My guess for the distance from flag C to flag D is _________ meters.

   The number of paces counted from flag C to flag D is _________ paces.

   My estimate of the distance from flag C to flag D is __________ meters.

4. My guess for the distance from flag D to flag E is _____ meters.

   The number of paces counted from flag D to flag E is _____ paces.

   My estimate of the distance from flag D to flag E is _____ meters.

5. My guess for the distance from flag E to flag A is _____ meters.

   The number of paces counted from flag E to flag A is _____ paces.

   My estimate of the distance from flag E to flag A is _____ meters.

6. When everyone has estimated the distances, use the tape measure to determine the actual distances between flags.

   The measured distance from flag B to flag C is _______ meters.

   The measured distance from flag C to flag D is _______ meters.

   The measured distance from flag D to flag E is _______ meters.

   The measured distance from flag E to flag A is _______ meters.

7. Were your distance estimates accurate within 2 meters?________

   Were they accurate within one meter? ________

   Did your estimates improve with practice? _________

   Why?____

8. Give an example of how you could use this distance-estimating skill.

   ____

   ____

9. What other quantities would you like to learn to estimate?

   ____

   ____

10. Why is the skill of estimation important? ______________________

   ____

   ____

Making the Grade

SUMMARY

Student topographic maps will be modified to represent the planned wetland. The changes in grade that must be made to accommodate the planned wetland will be shown in detailed cross sectional drawings.

OBJECTIVES

Students will decide the shape, size, and water depths of the planned wetland. Student topographic maps will be modified to reflect the grading changes that will be necessary to accomplish the wetland project goals. Students will create cross-sectional drawings of their planned wetland.

MATERIALS

- Completed *Topographic Map* Student Page
- Data on the volume of the planned wetland from the *Water Budget* Student Pages
- Completed *Wetland Goals* Student Page
- Copies of appropriate *Grading Plan* Student Page
- Copies of appropriate *Cross-sectional Views* Student Page

MAKING CONNECTIONS

Topographic maps show ground elevations as they exist. The grading plan is a representation of what the topography will be after construction. The grading plan is the link between the design idea and the actual project construction. The same is true of design plans for a soap box derby racer, a bird house, a garden, a road, or a city. Drawing plans and reading plans are lifelong skills.

BACKGROUND

Read Chapter 4, especially the section about *Grading Plans* (pages 59-61).

Grading is the process of changing the slope and elevation of the land. Before land is graded, it is first surveyed. Then a grading plan is developed that will accomplish the project goals. Grading is hard labor when done by hand; it is usually faster and easier if it is done with equipment. A grading plan prevents wasted time, effort, and money.

PROCEDURE

Warm-Up

Using pictures of bulldozers, excavators, and backhoes, query students as to what this equipment does and why. Has this equipment been seen excavating the basement of a building or preparing a road bed? Describe what was seen. Could the same work be done on a smaller scale with a shovel?

Activity

Grades K-4

If equipment is busy in the area, consider planning a trip to watch as workers move soil to change the contours of the land. Why is soil being moved from one place to another? Guess what will be built after the soil is moved.

Using modeling clay or sand in a sandbox, model how to change the land to form a wetland. Test it by adding water. Just as in real life, sandy soil may need a liner to prevent rapid infiltration (water soaking into the ground).

Grades 5-12

Students working in small groups will design a planned wetland using information from earlier activities. They will determine the functions of their planned wetland based on the *Wetland Goals,* the volume of water available using the *Water Budget,* and the placement of the wetland using the *Topographic Map.*

1. Within each group of students, a design should be discussed and a grading plan begun by penciling in new contour lines on a copy of the topographic map. Remember that wherever new contour lines are introduced, you must make sure that they connect to the same elevation somewhere on the plan. Even if only one contour line is modified, other lines must be adjusted to prevent steep slopes.

Designs might include an island on which birds and other animals might rest or nest safely, a berm (or other water control structure) to hold water at a chosen elevation, sloping edges to allow frog and turtle access, or a deep pool for fish. A walkway or observation deck might also be included if that feature meets the project goals and is financially possible. The design should clearly show how water will get into the wetland and how excess water will leave the wetland.

2. Trace the new contour lines onto the appropriate *Grading Plan* Student Page. Be sure to mark the benchmark used, the elevations of the contour lines, and the scale, as in **Figure 9.6**.

3. Draw one or more lines through the planned wetland on the *Grading Plan*; mark these lines A – A′, B – B′, etc. These lines will show the elevation changes along their lengths and are called cross sections. On the appropriate *Cross-sectional Views* Student Page, show the planned slopes and each planned elevation. See **Figure 9.7** for one example.

Pay attention to the steepness of the slopes. Slope is expressed as a ratio of horizontal distance to vertical rise. A 1:1 slope means that for every foot the slope extends horizontally, it rises one foot. A 1:1 slope is steep; actually it is a 45° angle. A 5:1 slope (18° angle) is much more gentle; for every five feet the slope extends horizontally, it only climbs or drops one foot vertically. By knowing the scale of your grading plan, you can accurately space your new contour lines to represent the desired slope. A slope that is no steeper than 3:1 is recommended for most wetland projects. Handicapped access usually requires a slope of 10:1 or greater.

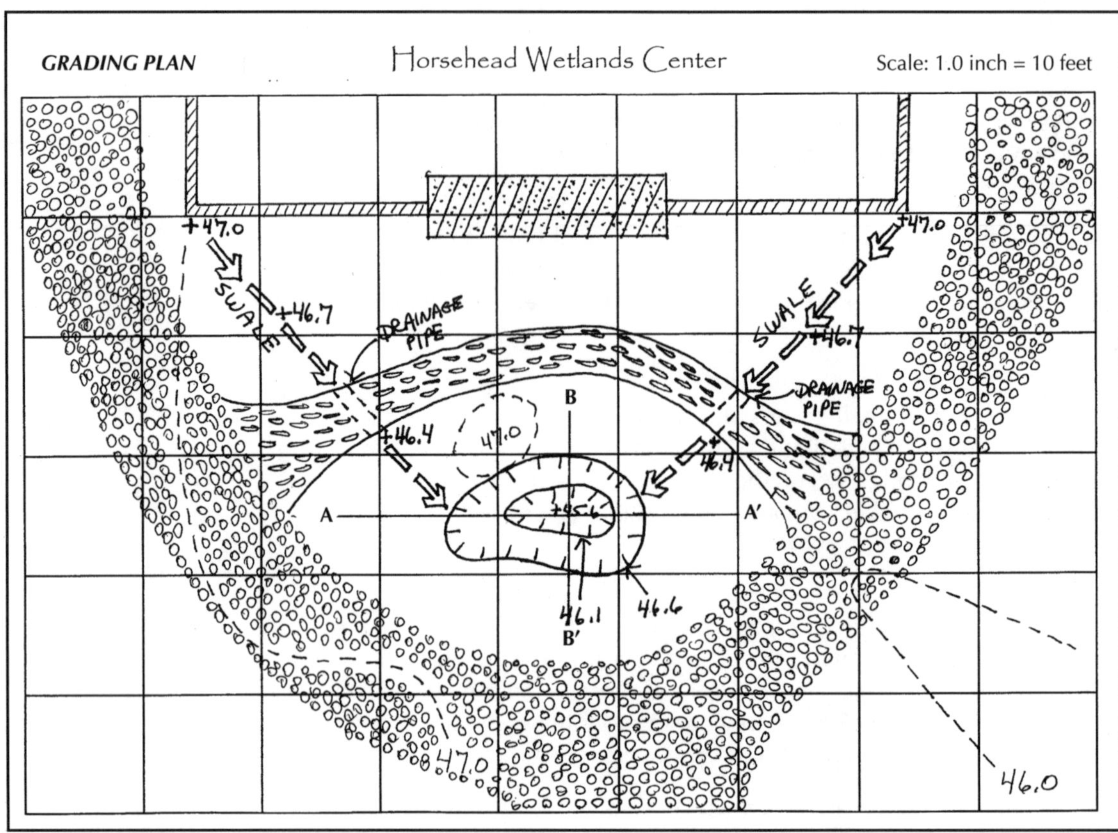

Figure 9.6 This is the *Grading Plan* created during the Horsehead Wetlands Center project.

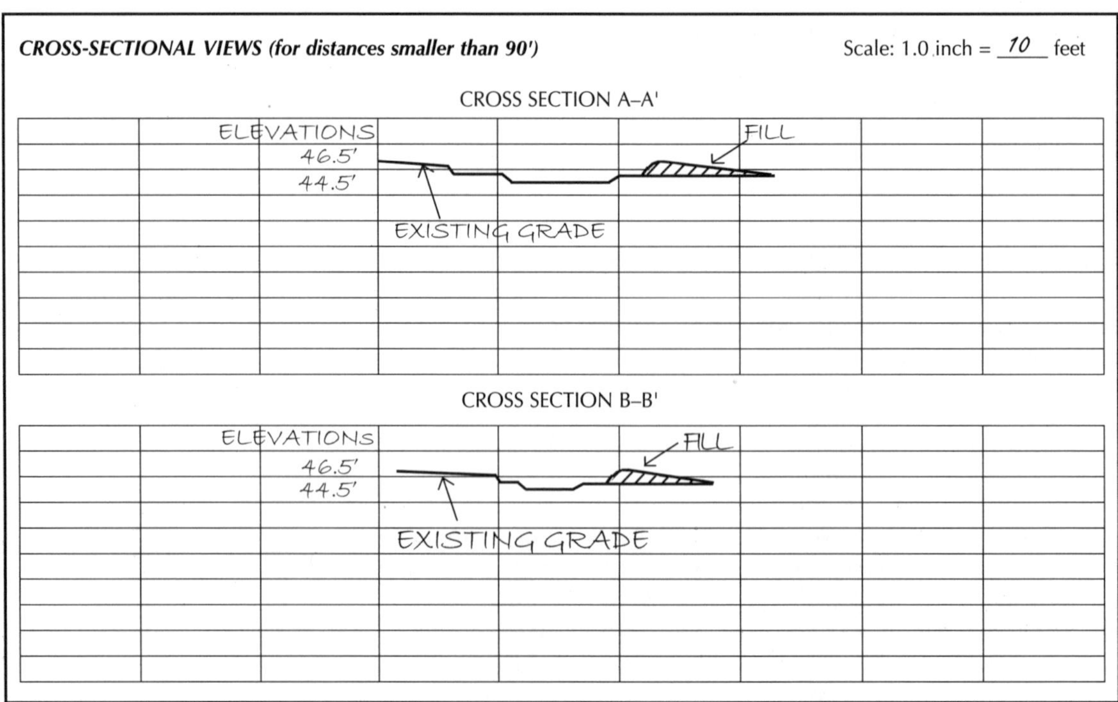

Figure 9.7 These are the *Cross-sectional Views* for the project at Horsehead Wetlands Center.

Wrap-Up

Compare the grading plans prepared by each group. Are they all alike? Are some more appropriate than others? As a class, select one grading plan to use for the planned wetland.

ASSESSMENT

Select the *Grading Plan* that best meets the following criteria:

- As designed, the wetland will hold the planned volume of water at the planned depth. Calculate the volume of the wetland to be sure.
- The grading plan meets the established wetland goals.
- The grading plan and cross-sectional views convey the information needed for others to do any necessary construction.
- Excess water has a means of leaving the wetland that will not cause problems.

EXTENSIONS

Locate grading plans for recently built roads, housing developments, shopping centers, or even schools. Compare current contours with the grading plans and regional topographic maps. How did the construction process change the contours of the land? Were the grading plans followed or modified?

RESOURCES

Reid, G.W. 1987. *Landscape Graphics.* Watson-Guptill Publications, New York, NY.

Strom, S. and K. Nathan. 1985. *Site Engineering for Landscape Architects.* The AVI Publishing Company, Inc., Westport, CN.

USDA-SCS. 1992. "Chapter 13: Wetland Restoration, Enhancement, or Creation", *Engineering Field Handbook.* U.S. Department of Agriculture, Soil Conservation Service, Washington, DC. (210-EFH, 1/92)

USEPA. 1996. *Protecting Natural Wetlands: A Guide to Stormwater Best Management Practices.* U.S. Environmental Protection Agency, Office of Water, Washington, DC. (EPA-843-B-96-001)

Wirshing, J.R. and R. H. Wirshing. 1985. *Schaum's Outlines: Introductory Surveying.* McGraw-Hill, Washington, DC.

GRADING PLAN (for areas smaller than 90' x 60')

Scale: 1.0 inch = ____ feet

GRADING PLAN (for large areas)

Scale: 0.5 inch = ______ feet

CROSS-SECTIONAL VIEWS (for distances smaller than 90')

Scale: 1.0 inch = _____ feet

CROSS SECTION A–A'

CROSS SECTION B–B'

CROSS-SECTIONAL VIEWS (for large areas)

Scale: 0.5 inch = _____ feet

CROSS SECTION A–A'

CROSS SECTION B–B'

Planting Plans

SUMMARY

Plant species will be selected for each planting zone based on regional appropriateness, availability, price, and the goals for the planned wetland.

OBJECTIVES

Students will compare plant characteristics and select those that satisfy the goals for their planned wetland. They must also specify which plants will be planted at each water depth (zone) and in each area of the planned wetland.

MATERIALS

- Plant catalogs from regional wetland nurseries
- Completed *Grading Plan* and *Cross-sectional Views* Student Pages
- Copies of appropriate *Planting Plan--Part I* Student Page
- Copies of *Planting Plan--Part II* Student Page
- Copies of *Order Form* Student Page
- Copies of *Wetlands* Student Page (if appropriate)

MAKING CONNECTIONS

Most people will plant a garden at some time in their lives. It may be as small as a few containers on a suburban apartment balcony or an urban rooftop. Maybe there will be an opportunity to plant a moderate-sized flower bed or vegetable garden. A few people will plant fields of flowers, vegetables, or trees. Whatever is planted, it will benefit both humans and surrounding creatures. Careful selection and placement of plants will result in a high rate of survival, and creatures that use those plants will be attracted to the area.

BACKGROUND

Read Chapter 4, especially the section on *Plans and Specifications* (pages 61-64).

Each species of plant is adapted to a specific hydrologic regime; this includes the frequency and depth of flooding, the timing of floods and droughts, the salinity of the water, and sometimes the pH of the water. For each of these factors there is an **optimum range** at which each plant species can grow and a broader **range of tolerance** at which each species can survive. If plants are placed in areas to which they are adapted, they are more likely to survive, flourish, and reproduce.

Plants that are selected for a planned wetland should reflect the project goals. Benefits that a planned wetland might provide include: frog habitat, bird nesting areas, aesthetics, water quality enhancement, sediment stabilization, etc. If attracting wildlife for nature studies is a project goal, this may be accomplished through the wetland design and the selection of plant species. Amphibians may be attracted to the water during their breeding seasons. Birds, butterflies, insects, and mammals will be attracted to specific seeds, nectar, or foliage. Predatory creatures will be attracted because of prey animals that feed and hide in the wetland.

Selection of plant species native to the area, instead of nonnatives or even invasive species, will help students recognize the natural ecosystem as it existed before human interference. This can lead to lessons in the historical use of plants by earlier occupants of the land. Native plants also require less care once established, and support animal species that are native to the area.

PROCEDURE

Warm-Up

Show students pictures of at least three local native wetland plant species from a local or regional magazine or a wetland poster. Have students identify the plants and suggest ways that these plants are used by wetland creatures.

Activity

Grades K-4

Explain the difference between a tree (single trunk, usually more that 20 feet tall), a shrub (many stems, less than 20 feet tall), an emergent plant (stands in water), and a submerged plant (usually under water, may have floating leaves). Locate examples on a wetland poster or picture.

On the *Wetlands* Student Page, students color and cut out the wetland plants, then attach them at an appropriate location on the wetland drawing. Several arrangements are possible, but the coontail must be underwater.

Grades 5-12

1. Using copies of the *Grading Plan*, outline the following zones on the appropriate *Planting Plan--Part I* Student Page:

Zone	Water	Types of Plants
Upland	None	Trees and shrubs
Transitional	Soggy soils	Herbaceous emergent plants and shrubs
Shallow wetland	0-6" water	Herbaceous emergent plants
Deep wetland	6-18" water	Herbaceous emergent plants
Pond	≥18" water	Floating leaved plants and submerged aquatic vegetation

2. Working in small groups, select three plant species and one substitute for each zone. (Each group may be assigned a different zone.) Each selected species should be adapted to (prefer) the hydrology, shade, salinity, and pH of that zone. For each plant species selected, also determine if the habitat and intended benefits match the goals and design for the planned wetland. (See Chapter 7, *Common Wetland Plants*.) If the plants selected are not suitable, choose other plants. Record selections on the *Planting Plan--Part II* Student Page as in **Figure 9.8**.

PLANTING PLAN--PART II

Plant Species	Water Tolerance	Shade Tolerance	Height Range	Flowering Time	Intended Benefits
A. Red osier dogwood	seasonal flooding	partial	6' to 12'	May to June	wildlife food and cover
B. Witch hazel	irregular flooding	full	20' to 30'	Sept. to Dec.	beauty
C. Black-eyed susan	seasonal flooding	partial	1' to 3'	Aug. to Oct.	beauty
D. New York aster	irregular flooding	none	1' to 3'	July to Oct.	beauty and cover
E. Switch grass	seasonal flooding	none	2' to 4'	July to Sept.	winter beauty, wildlife food
F. Black chokeberry	irregular flooding	partial	6' to 10'	May	songbird food
G. Little blue stem	seasonal flooding	none	2' to 3'		wildlife cover
H. Marsh hibiscus	0 to 3" water	none	4' to 7'	July to Sept.	hummingbird & wildlife food
J. Tussock sedge	0 to 6" water	none	2' to 4'	May to Aug.	wildlife food and cover
K. Swamp milkweed	seasonal flooding	none	3' to 6'	June to Aug.	butterfly nectar
L. Joe-Pye weed	seasonal flooding	partial	2' to 5'	July to Sept.	butterfly nectar
M. Cardinal flower	irregular flooding	none	2' to 5'	July to Oct.	hummingbird & butterfly
ALTERNATIVES:					
Blue flag	regular flooding	partial	1' to 3'	Late spring	beauty
Arrowwood	seasonal flooding	partial	6' to 12'	May to June	bird and mammal food

Figure 9.8 This is the *Planting Plan--Part II* created for Horsehead Wetlands Center.

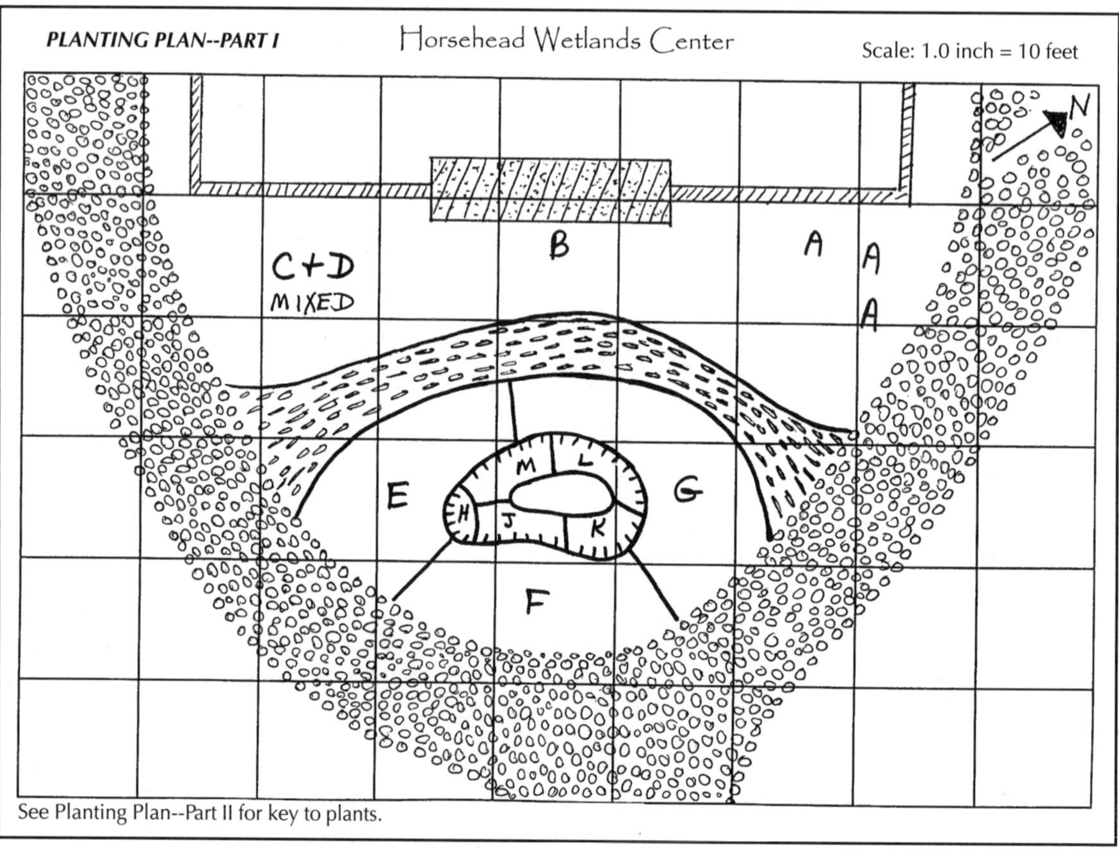

Figure 9.9 This is the completed *Planting Plan--Part I* for Horsehead Wetlands Center.

3. Each group should determine where the species selected will be placed within the planting zone and indicate the locations on the *Planting Plan--Part I* as in **Figure 9.9**. During design of the planting plan, use a key to mark plant species' locations. Hint: clustering the same species of plants together gives a more natural appearance to the planned wetland and makes it simpler to plant and monitor.

4. Using a local wetland nursery catalog, enter the following information on the *Order Form* Student Page:

a. the **Rate of Spread**: slow, medium, or fast;
b. the **Spacing** or distance apart that plants should be place: one foot apart if slow spreading, two feet apart for medium rate of spread, three feet apart for those that spread rapidly, and five to twenty feet apart for trees and shrubs;
c. the **Planting Area** per species, which is determined using the same technique as on page 164;
d. the **Number** of plants needed to cover an area, which is calculated by dividing the **Planting Area** by the **Spacing** squared;
e. the **Form** in which the plant species is available: pots, bare roots, plugs;
f. the **Cost Per Plant**, which will vary by nursery and the form available;
g. the **Total Cost Per Species**, which is calculated by multiplying the **Number** times the **Cost per Plant.**

To determine the **Total price for plants** for the planned wetland, add the amounts in the **Total Cost Per Species** column, but do not include substitutes.

Wrap-Up
As a class compare planting plans based on the appropriateness of the species selected for each zone, the goals that will be met (i.e., plants to attract butterflies, nesting for birds), and overall cost. Select one planting plan to be implemented.

ASSESSMENT
Review the *Planting Plan--Part I,* the *Planting Plan--Part II,* and the *Order Form* prepared by each group of students.

- Are the selected plant species appropriate?
- Is a suitable key provided?
- Are the project goals likely to be met by the plants that were selected?
- Are calculations of plant quantities and costs accurate and within the budget limits?

EXTENSIONS
Choose one of the following situations. Suggest reasons for what has happened and design a second planting plan.

a. All of the plants survive for two years after the planned wetland is planted.

b. Throughout the wetland, half of the plants die during the first two years after the planned wetland is planted.
c. All of the plants die in one area of the planned wetland.
d. All of one species of plant die throughout the entire planned wetland.

RESOURCES

Cox, J. 1991. *Landscaping with Nature*. Rodale Press, Emmaus, PA.

Druse, K. 1994. *The Natural Habitat Garden*. Clarkson Potter Publishers, New York, NY.

Ottesen, C. 1995. *The Native Plant Primer.* Harmony Books, New York, NY.

Perry, F. 1981. *The Water Garden.* Van Nostrand Reinhold Company, New York, NY.

Seidenberg, C. 1995. *The Wildlife Garden.* University Press of Mississippi, Jackson, MS.

Stevenson, V. 1985. *The Wild Garden.* Penguin Handbooks, New York, NY.

Tufts, C. and P. Loewer. 1995. *Gardening for Wildlife.* Rodale Press, Emmaus, PA.

USEPA. 1994. *A Citizen's Guide to Wetland Restoration.* U.S. Environmental Protection Agency, Region 10, Seattle, WA. EPA 910/R-94006.

Black-eyed susan,
Rudbeckia fulgida

WETLANDS (for grades K-4)

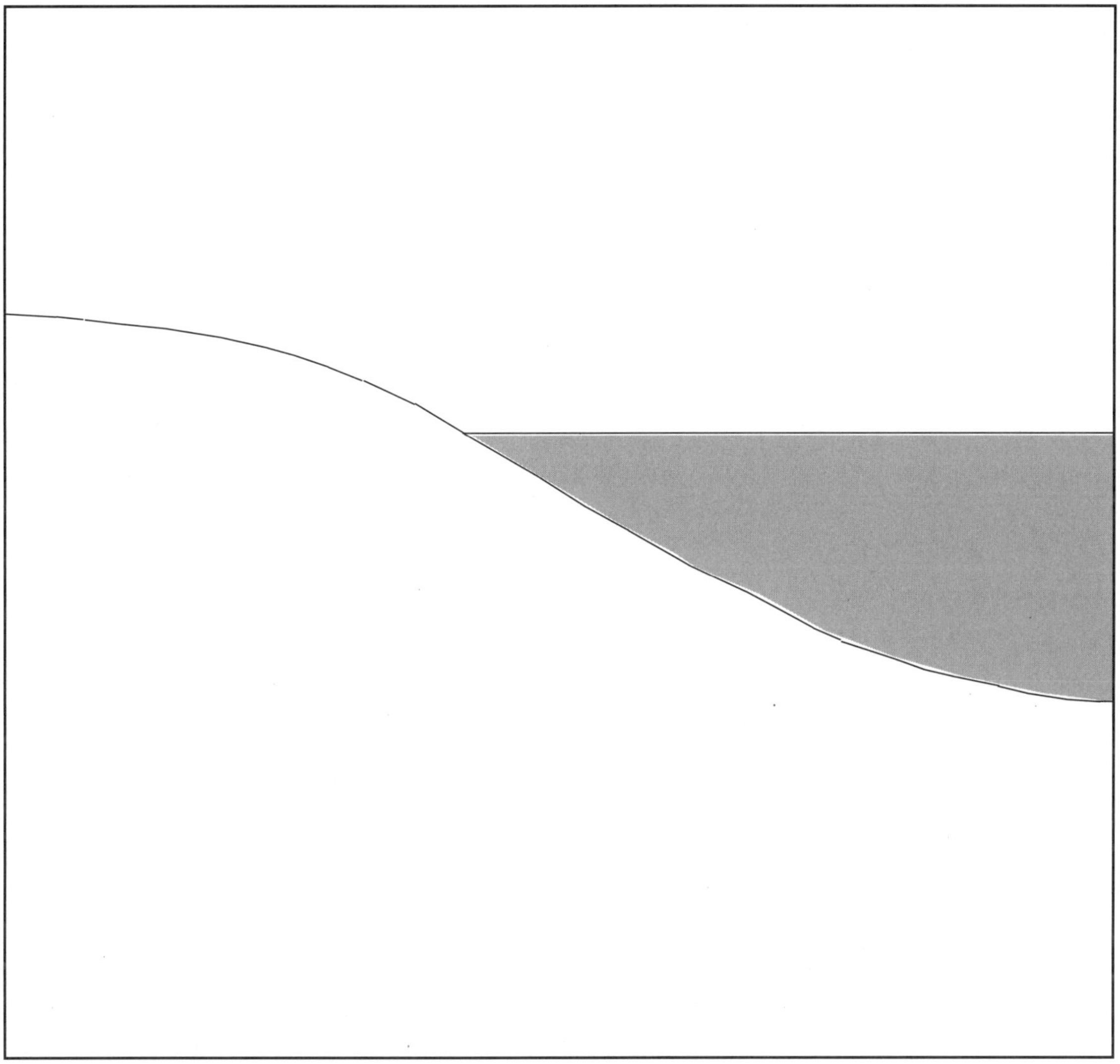

Cut out the plant squares and place them where they belong in the wetland.

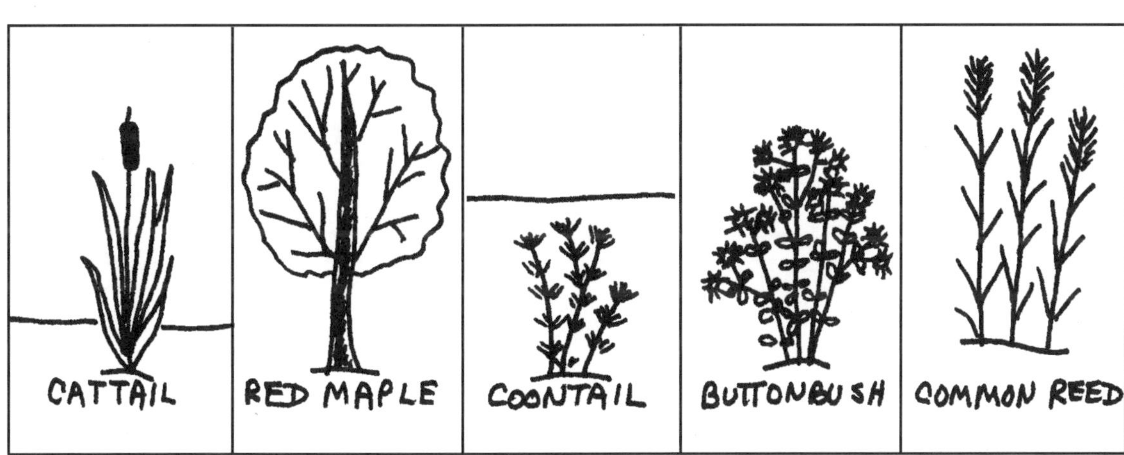

PLANTING PLAN--PART I (for areas smaller than 90' x 60')

Scale: 1.0 inch = _____ feet

PLANTING PLAN--PART I (for large areas)

Scale: 0.5 inch ______ feet

PLANTING PLAN--PART II

Plant Species	Water Tolerance	Shade Tolerance	Height Range	Flowering Time	Intended Benefits

ORDER FORM

Plant Species	Rate of Spread	Spacing	Planting Area	Number	Form	Cost Per Plant	Total Cost Per Species
					Total price for plants:		

Clarifying Specs

SUMMARY

Grading and planting plans will be examined in detail, and specifications will be written for the planned wetland.

OBJECTIVES

Students will determine grading and planting details of the planned wetland in the form of specifications that will be used for permitting (if necessary), construction, and planting. A construction sequence and timetable will be established within the specifications.

MATERIALS

- Completed *Wetland Goals* Student Page
- Completed *Grading Plan* and *Cross-sectional Views* Student Pages
- Completed *Planting Plan--Part I* and *Part II* Student Pages
- Copies of *Grading Specifications* Student Pages
- Copies of *Planting Specifications* Student Pages

MAKING CONNECTIONS

Prior to any type of construction (houses, roads, shopping centers, etc.), plans are drawn and specifications written with sufficient detail provided to those doing the work, for the end result to match the vision of the planners. This is a time when problems and omissions are discovered and rectified with minimal effort and expense. Plans and specifications are necessary for permits and are the basis for a performance contract if an outside contractor is hired.

BACKGROUND

Read Chapter 4, especially the section about *Plans and Specifications* (pages 61-64).

In an earlier activity a grading plan for the planned wetland was developed. Now details must be specified before permits (if needed) can be issued, contracts (if necessary) can be signed, and construction can begin. Aspects of the planned wetland that may require additional information include type, method of installation, and cost of liners; type, method of installation, size, location, and cost of water control devices; and costs for excavation equipment should it be needed. A location for stockpiling or disposing of excess excavated material must also be determined.

Some soils allow infiltration at too fast a rate to hold water. To create a wetland under these circumstances, the soil must be compacted and additional water supplied during the first year or a liner purchased and installed. Many types of liners are available at a range of prices.

Water control devices may be necessary for water coming into the wetland (intake) and/or water leaving the wetland (discharge). Types and prices vary according to the design and size.

Excavation equipment might be as simple as a rented roto-tiller to break up (deconsolidate) the soil and shovels for students to dig, or it could be full-sized construction equipment with trained operators. Prices vary considerably. What is needed will depend on the size and design of the wetland, and the physical abilities of students and volunteers.

Plants may need fertilizer to help them become established. Solid fertilizers that do not float and do not promote algal growth by getting into the water are recommended. Other soil amendments may be necessary to improve soil texture for healthy plant growth.

PROCEDURE

Warm-Up

Since most wetland goals include habitat for wildlife, imagine that you are a frog. If you could have a wetland built for you, what would it be like? (Wet during the breeding season.) How big would the wetland be? (Big enough to hold water while eggs hatch and tadpoles develop into frogs.) Do you need water all the time, or just sometimes? (Sometimes, just long enough for my eggs to hatch and the tadpoles to develop into frogs.) If you need water, how deep should it be? (A few inches deep; if it is deep enough for fish, they will eat my offspring.) What special features would you like? (Places to hide from birds, snakes, and other animals that could eat me. Lots of plants for the tadpoles to eat, and insects for me to eat.)

Activity

Grades K-4

Have pairs of students pretend to be ducks, snails, swallows, beavers, butterflies, dragonflies, minnows, herons, raccoons, snakes, water turtles, water beetles, and other wetland creatures. Find out where these creatures hide, what they eat, how much water they need, and where they raise their young. Do any of the creatures have special needs, such as a place to sun themselves?

Help students make a mural of the planned wetland and plants. Students may draw or cut out pictures of their creature, then find a place in the wetland mural where their creature would be happy. Be sure each creature has food, water, and shelter!

Students may now act as their creatures might in an informal play in front of the mural or, better yet, write a wetland play of their own!

Once the planned wetland it complete, students can verify that the needs of "their" creatures are met by carefully inspecting the wetland.

Grades 5-12

A. Working in small groups, review the *Grading Plan*. Make sure the following items are present on the grading plan and that design of the wetland meets the goals expressed in *Wetland Goals*.

1. All planned water levels are clearly shown.
2. The bottom elevations (or invert) of all water-control structures (both intake and discharge) are shown.

3. The location and elevation project benchmarks are clearly indicated.
4. One foot contour lines are visible.
5. All roads, property lines, buildings, and other structures are clearly marked.
6. *Cross-sectional Views* show all elevation changes to the planned wetland, such as shelves, islands, and berms.

B. In small groups, complete the *Grading Specifications* Student Pages. If more information is needed, check written information available in the library, locate internet resources, or contact professionals. This may be a good time to invite a guest speaker (someone who works in construction or planning) to answer student questions. Be sure to address safety issues such as the possible need for fencing.

C. Using information from the *Planting Plan--Part I* and *Planting Plan--Part II,* complete the *Planting Specifications* Student Pages. Give careful consideration to management of invasive species of plants in the local area, unwanted attention from animals (such as deer, geese, or muskrats browsing on new plants), and whether seasonal mowing will be appropriate. Remember that native grasses provide much needed shelter during winter. Types and amounts of fertilizer, and whether or not to add soil amendments (such as sand or mulch) are other considerations. A guest speaker from a plant nursery or garden club might be helpful in answering questions.

Wrap-Up

As a class, compare *Grading Specifications* and *Planting Specifications* based on completeness, the goals that will be met (for example deep open water for fish, a bank or log for turtles to sun themselves), and overall cost. Select the most appropriate *Grading Specifications* and *Planting Specifications* to be used for permits, contracts, construction, and planting of the planned wetland.

The *Grading Plan, Cross-sectional Views, Grading Specifications, Planting Plans,* and *Planting Specifications* are used for actual construction and planting of the planned wetland, and as such will become a permanent record of the planned wetland.

ASSESSMENT

Review the *Grading Specifications* and *Planting Specifications* prepared by each group.

- Are the goals expressed in *Wetland Goals* likely to be met?
- Are the *Grading Specifications* complete and appropriate?
- Are the *Planting Specifications* complete and appropriate?

EXTENSION

Modify the current plans to include one or more of the following:

a. a boardwalk or bridge,
b. a deep-water pool for fish,
c. an island to protect birds from predators,
d. a sloped bank for easy access by reptiles and amphibians,
e. plants that will attract butterflies.

RESOURCES

Denbow, T.J., D. Klements, D.W. Rothman, E.W. Garbisch, C.C. Bartoldus, M.L. Kraus, D.R. Maclean, and G.A. Thunhorst. 1996. *Guidelines for Development of Wetland Replacement Areas.* National Cooperative Highway Research Program Report 379. National Academy Press, Washington, DC.

Husted, R. 1997. *Wetlands for Clean Water.* Clean Water Network and Natural Resources Defense Council. Washington, DC.

Interagency Workgroup on Wetland Restoration. 1999 DRAFT. *Interagency Federal Guidance on Wetland Restoration, Creation, and Enhancement.* USEPA, NOAA, ACOE, FWS, NRCS, Washington, DC.

Kusler, J.A. and M.E. Kentula, editors. 1990. *Wetland Creation and Restoration, The Status of the Science.* Island Press, Washington, DC.

Lewis, R.R. 1982. *Creation and Restoration of Coastal Plant Communities.* CRC Press, Boca Raton, FL.

For information on liners contact: CETCO, Lining Technologies Group, 1350 West Shure Drive, Arlington Heights, IL 60004-1440; 847-392-5800.

For local resources: The local Natural Resource Conservation Service office and the Soil Conservation Service office can provide detailed information on local soils and hydrology, while the county Extension Service office has information on native plants and local pests. Both are excellent resources, providing both advice and speakers. Other local resources include nurseries that construct water gardens; and the engineering departments of governments, landfills, sewerage treatment plants, electric utilities, or mines.

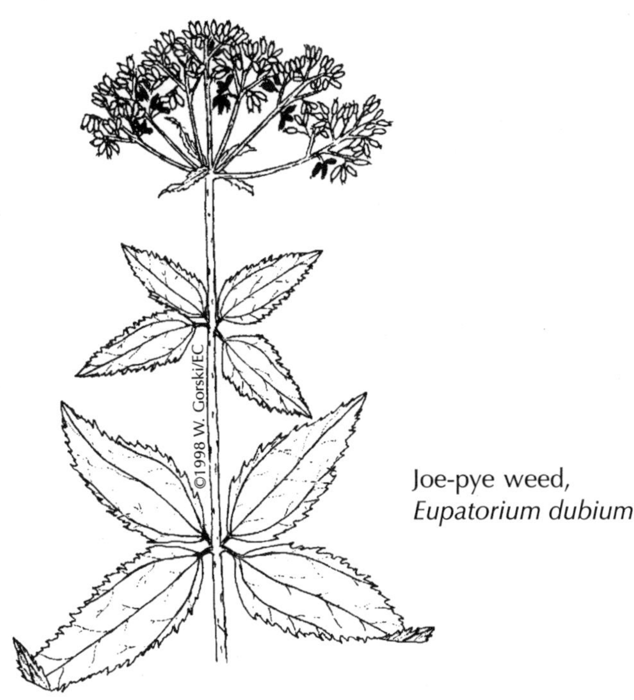

Joe-pye weed,
Eupatorium dubium

GRADING SPECIFICATIONS

1. Construction sequence and timetable:

Construction Activity	Date Planned	To Be Done By Whom
Materials purchased		
Material delivery date(s)		
Utility lines located and flagged		
Equipment on site (shovels, roto-tiller, heavy equipment)		
Clearing of site and grubbing of roots		
Deconsolidation of soils		
Staking of excavation site		
Excavation of wetland		
Construction of berms, islands, etc.		
Installation of water control devices		
Installation of liner		
Plants purchased		
Plant delivery date(s)		
Flagging of planting areas		
Planting date(s)		
Seeding date(s)		
Construction of walkways, observation blinds, etc.		
Other		

2. Special considerations and conditions (such as erosion control during construction of large wetlands or existence of a 30 inch doorway through which equipment must fit):

3. Volume of material to be excavated:

4. Excavated material will be:

    a. stockpiled (give location and time frame)

    b. disposed of (where and how)

    c. used on site (where)

5. Details of materials:

Material	Details	Quantity
Liner		
Water-control structure		
Fence and gates		

6. What labor will be done by:

    a. contractors

    b. volunteers

    c. students

7. Which materials will be:

    a. purchased

    b. donated

    c. borrowed

PLANTING SPECIFICATIONS

1. Plant species and planting zones:

Water Depth Range	Area in Square Feet	Plant Species	Spacing	Form of Plant Material	Number of Plants

2. Areas to be seeded:

   a. Species in seed mixture and percentage composition of each:

   b. Seed at a rate of _________ lbs/acre.

3. Acceptable substitutes for plant species, should some not be available:

4. Planting details:
   a. Attach *Planting Plan--Part I.*
   b. Types and amounts of fertilizers to be used:
   c. Types and amounts of soil amendments to be used:

5. Seeding details:

6. Watering requirements:

7. Details of wildlife control structures (if needed):

8. Maintenance needs:

10 MONITORING A WETLAND

ACTIVITIES

Faithful Flora

Floral Friends

Soggy Soils

Healthy Water

Marvelous Hydrology

Data Sandwiches

©2000 C. Barcomb/EC

Faithful Flora

SUMMARY

Wetland plants will be monitored by measuring plant survival, stem diameter, basal area, plant height, amount of cover, and biomass.

OBJECTIVES

Properties of individual plants and plant populations will be examined and evaluated to assess the success and health of the wetland.

MATERIALS

- Plant identification field guides and keys
- Hoops, approximately one meter in diameter
- Survey flags
- Sticks
- Camera
- Photo album
- Rulers
- String or tape measures
- Meter sticks
- Drying oven
- Brown paper bags
- Balances for measuring biomass
- Copies of *Individual Plants* Student Page
- Copies of *Plant Summary* Student Page

MAKING CONNECTIONS

As with a garden that has been planted, a planned wetland is successful when its plants survive, grow, and reproduce.

BACKGROUND

Wetland plants have adapted to conditions caused by the presence of water. Plant roots need oxygen for respiration and growth. Consequently, wetland plants must get their roots to oxygen or oxygen to their roots under the anoxic (without oxygen) conditions present in shallow water and wetland soils. In the presence of water, some insoluble (oxidized) elements in the soil change by oxidation-reduction reactions into soluble (reduced) forms that are readily taken up by plants. Because of this, wetland plants must adapt to coping with levels of iron, manganese, sulfur, and nitrogen that may be toxic to other plant species.

The survival of individual plants and/or species of plants is one measure of a wetland's success. Growth rates as determined by changes in stem diameter, plant height, or degree of cover over time are other measures of the success of a wetland. Determining plant productivity by measuring biomass is a more destructive means of evaluating success, but it may be useful in making comparisons in experimental situations.

PROCEDURE

Warm-Up

Show an artificial flower to the group. What characteristics indicate that it is not alive? Show a living plant and a dried flower to the group. List characteristics of individual living plants and dead plants. List and discuss shared ideas on the board. Come to a consensus on what criteria will be used to determine which wetland plants are alive and which have died.

Activity

With students working in pairs or small groups, try the assessment techniques that seem most appropriate. Suggested grade levels are indicated, with the activities presented in order of increasing skill level. Field guides should be used as needed to identify plant species.

Usually monitoring all plants within a wetland is too difficult a task to accomplish unless the wetland is small. Transects and quadrants provide smaller areas (subsamples) that can be more thoroughly examined. The data gained from the smaller samples can be extrapolated to represent the entire wetland. Several transects throughout the wetland may be established as in the activity *What Grows There?* on page 127. Another option is to mark several scattered one-meter square (or larger) quadrants throughout the wetland using sticks and string, then analyze the plants within the quadrant. Or, if the vegetation is short, a hoop can be tossed randomly into the wetland, and plants within the hoop can be assessed.

In each case, be sure to measure and calculate the area sampled. For a transect or quadrant, use the formula:

$$\textbf{area} = \textbf{length} \times \textbf{width}.$$

For a hoop use the formula:

$$\textbf{area} = \pi r^2, \text{ if the radius is measured, or}$$

$$\textbf{area} = \pi (d/2)^2, \text{ if the diameter is measured.}$$

<u>Grades K-12</u>

A. PHOTOGRAPHIC MONITORING

Photographic monitoring of the entire wetland community is an effective tool for assessing survival. Choose several locations around the wetland to be used as permanent photographic sites. The number of locations depends on the wetland size. Mark these locations with a colored wooden stake, plastic pipe, a large stone, a brick, or some other type of marker that will last for a number of years and will not be moved. If photographs are taken at each site each season on approximately the same date (fall, winter, spring, summer), annual comparison will provide visual evidence of changes occurring in the wetland. Have the plants increased in size? Is there more or less ground covered by plants? Have some species of plants been replaced by invasive species, like common reed, purple loosestrife, or even cattails?

Maintain a photo album of the photographs for yearly comparisons and assessment of any changes. Be sure each photograph is clearly marked with its date and location.

B. PLANT SURVIVAL, PART I
Plants planted in the wetland can be monitored for survival by using the planting plans to locate them, then assessing whether they are alive or dead. Of course, this is only possible during the growing season. Why? (The above-ground parts of many plants appear dead during winter, but the below-ground parts are probably alive.) Record the number of **Live** plants and the number of **Dead** plants of each species on the *Plant Summary* Student Page.

Grades 5-12

C. PLANT SURVIVAL, PART II
Make the following calculations and record the percent survival for the planted wetland plants on the last line of the *Plant Summary* Student Page.

a. Total number of dead plants = ?

b. Total number of live plants = ?

c. Total number of plants planted = number of live plants + number of dead plants = ?

d. Percent loss = number of dead plants ÷ total number of plants = ?

e. Percent survival = number of live plants ÷ total number of plants = ?

f. If some plants died, were they all of the same species? If so, what might have caused the death of this species? Was this species planted at an appropriate location relative to the water level of the wetland?

g. Loss of up to 15% of new plants is acceptable. Were your losses greater? If yes, what might be the cause? Was supplemental water supplied during the first dry season?

h. Have insects, diseases, or herbivores attacked the plants? Plants should be monitored for the presence of disease and attack by such voracious herbivores as geese, deer, nutria, and muskrat. (See *The Dynamics of Wetlands* on pages 24-30.)

D. STEM DIAMETER OR BASAL AREA
Stem diameter is the maximum width of a plant stem. The **Stem Diameter** of most plants can be measured using a ruler and without cutting or injuring the plant. The measurement should be made one inch (2.5 cm) above ground level, because stem diameters usually become smaller higher up on the plant. For trees, the measurement should be made 4.5 feet (1.4 meters) above ground level; this is known as the diameter at breast height, or d.b.h. Be sure to record all measurements on the *Individual Plants* Student Page.

Calculate the **Average Stem Diameter** per plant species by adding all the widths and dividing by the number of plants of that species. Calculate the **Area Sampled** by multiplying the length of the area sampled times the width of the area sampled. Record the calculations on the *Plant Summary* Student Page.

Basal area is the area (A) of the surface of the tree's stump if a tree were to be cut at breast height. (Measuring the basal area of a tree is preliminary to determining how much of the tree can be used as lumber.) To calculate basal area of small trees, when the diameter can be measured, use the formula:

$$A = \pi (d/2)^2, \text{ where } d = \text{diameter and } \pi = 3.14.$$

For large trees, measure the circumference 4.5 feet above ground level using a tape measure or string and meter stick. Calculate the **Basal Area** for each tree using the formula:

$$A = c^2 \div 4\pi, \text{ where } c = \text{circumference and } \pi = 3.14.$$

Record the **Basal Area** on the *Individual Plants* Student Page.

Calculate the **Average Basal Area** per tree species by adding all the basal area numbers for one species and dividing by the number of trees of that species. For a quadrant, calculate the **Area Sampled** by multiplying the length of the area sampled times the width of the area sampled. Be sure to record these calculations on the *Plant Summary* Student Page.

E. PLANT HEIGHT

Plant height is the distance from the ground at the base of the plant to the very top of the plant. **Plant Height** (H_t) can usually be measured directly using a meter stick. Record measurements on the *Individual Plants* Student Page.

The size of shadows on a sunny day can be utilized in a ratio to determine the height of tall shrubs and trees. Students must measure their own height, the length of their shadow, and the length of the tree's shadow. The height of each tree (H_t) will equal the student's height (H_s) times the length of the tree's shadow (L_t) divided by the length of the student's shadow (L_s).

$$H_t = H_s \times L_t \div L_s$$

If several trees are to be measured, have students measure all of the tree shadows and one student shadow at the same time. Why would this be important? (Shadow length changes over time; so measuring all shadows at the same time allows greater accuracy.)

Now calculate the **Average Height** per species by adding all of the heights together and dividing by the number of plants. Also calculate the **Area Sampled** by multiplying the length of the area sampled times the width of the area sampled. Record this data on the *Plant Summary* Student Page for comparison in following years.

F. PLANT COVER

Cover is the area of ground covered by a plant's shadow at noon. This is called canopy cover if tall shrubs and trees are being measured. Measure the widest diameter (d) of the plant's shadow at noon, then calculate the area (A) using the formula:

$$A = \pi (d/2)^2$$

Measurements need not be made at noon, but may be estimated at any time by looking down at smaller plants (see **Figure 10.1a**) or up at the canopy of taller plants, then placing small sticks in the ground to mark the area that would be covered by the plant's shadow at noon (see **Figure 10.1b**). Ignore small gaps between branches. Record the diameter measurements (**d**) and calculations of the area covered (**A**) on the *Individual Plants* Student Page. Calculate the **Average Area of Cover** for each species by adding the areas and dividing by the number of plants. Calculate the **Area Sampled** by multiplying the length of the area sampled by the width of the area sampled, then record the data on the *Plant Summary* Student Page for comparison in future years.

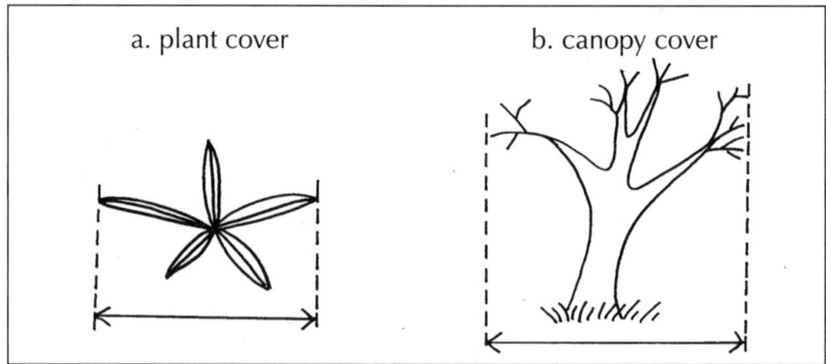

Figure 10.1 (a) Looking down on shorter plants, measure the diameter of the area that would be shaded at noon. **(b)** Looking up at the taller plants, mark on the ground where the outermost edges of the shadow would be if the sun were directly overhead at noon. Measure the diameter of the area marked on the ground.

Grades 7-12

G. PLANT BIOMASS

Biomass is the dry weight of part or all of a plant. The process used to determine biomass is a destructive process that should only be used for specific plant parts or representative plants (for example, three cattails out of a stand of thirty). Depending on the species of plant and the comparisons to be made, biomass can be measured for the above-ground parts, the below-ground parts, flowers, seeds, leaves, or stems.

First collect all of the plant parts or plants to be evaluated. Measure the **Wet Plant Biomass** for each individual plant (or the parts from a plant). Use a measuring balance that is appropriate for the mass and size of the plant. Tall plants may be cut into pieces, each piece weighed, then the weight of the pieces totaled. Place each plant (or its parts) in a brown paper bag and label it. Dry them very slowly in a warm (not hot) oven, until their mass is the same for two consecutive dry measurements. Record the **Dry Plant Biomass** on the *Individual Plants* Student Page. Calculate the **Average Biomass Per Plant** for each species by adding the biomass and dividing by the number of plants. Record the information on the *Plant Summary* Student Page.

Wrap-Up

Store all the data collected in this activity to compare with previous and future years. Our understanding of wetlands is evolving based on knowledge gained from such accumulated data.

What changes are occurring to the wetland plants from season to season, and year to year? Describe these changes and hypothesize reasons for them.

ASSESSMENT

Have students do the following:

- Draw or name three to ten plant species found in the wetland and list their characteristics.
- Orally or in writing, name and describe the tallest and shortest plants, the widest and thinnest plants, the plants with the greatest and least cover, and those with the greatest and least biomass.
- Is the wetland successful and healthy? What data supports this evaluation?
- Compare the most recent data collected with previous data and give reasons for the changes that have occurred in the wetland over time.
- Hypothesize how the wetland plants will change in the future, giving reasons for the changes.

EXTENSIONS

If this is the first year for the activity, have students create separate line graphs of average stem diameter, average basal area, average height, average area covered, and average biomass/plant. Set up the graphs for a ten year period, but only enter the data points for the first year. Five to ten species can be plotted on each graph if different colors or symbols are used for the data points.

If this is not the first year for the activity, have students add their data points to the graphs that have already been created.

Create a field guide, illustrating the most abundant plant species and listing their characteristics.

RESOURCES

Caduto, M.J. 1990. *Pond and Brook.* University Press of New England, Hanover, NH.

Firehock, K., L. Graff, J.V. Middleton, K.D. Starinchak, and C. Williams. 1998. *Save Our Streams: Handbook for Wetlands Conservation and Sustainability.* Izaak Walton League of America, Gaithersburg, MD.

Kent, D.M. 1994. *Applied Wetlands Science and Technology.* CRC Press, Boca Raton, FL.

Project Learning Tree, Environmental Education Activity Guide. 1995. American Forest Foundation, Washington, DC. Especially: Trees in Trouble and How Big Is Your Tree?

Tiner, R.W. 1998. *In Search of Swampland: A Wetland Sourcebook and Field Guide.* Rutgers University Press, New Brunswick, NJ.

INDIVIDUAL PLANTS

DATE: ____________________

Plant Species	Stem Diameter	Basal Area *	Plant Height *				Plant Cover		Plant Biomass	
	d	A	H_s	L_t	L_s	H_t	d	A	Wet	Dry

* Shaded columns are for large tree measurements only.

PLANT SUMMARY

DATE: ____________

Plant Species	Survival		Average Stem Diameter	Average Basal Area *	Average Height	Average Area of Cover	Average Biomass Per Plant	Area Sampled
	Live	Dead						

* Shaded column is for tree measurements only.

Floral Friends

SUMMARY

Plant community cover type, percent vegetative cover, density, and productivity may be determined. Species richness, abundance, and frequency within the plant community may also be evaluated.

OBJECTIVES

Properties of plant species and of the entire plant community will be examined and evaluated to assess the success and health of the wetland.

MATERIALS

- Plant identification field guides and keys
- Poster *Wetlands: Water, Wildlife, Plants and People!*
- Hoops, approximately one meter in diameter
- Completed copies of *Individual Plants* and *Plant Summary* Student Pages
- Copies of *Plant Community Summary* Student Page
- Copies of *Species Abundance Graph* Student Page

MAKING CONNECTIONS

As with any garden, the success of a planted wetland can be measured by the survival, growth, and reproduction of its plants. If the hydrology is right, the plants will survive; and once the plants are established, the animals will come.

If weeds and/or invasive species are also controlled or eliminated, then diversity of both plants and animals can be maintained. If only a few plant species are present in a wetland (as in a monoculture like a cornfield), then the vegetation is more susceptible to pests and less able to adapt to changing conditions.

BACKGROUND

Determination of wetland community success and change over time can be assessed by methods such as percent vegetative cover, cover type, density, productivity, species richness, species abundance, species frequency and distribution.

PROCEDURE

Warm-Up

Working in small groups, select a particular wetland to assess on the U.S. Geological Survey poster *Wetlands: Water, Wildlife, Plants and People!* Have students count the number of cattails (or trees) in the wetland selected. If each centimeter on the poster represents one meter on the ground, measure and calculate the area of the wetland (length times width in meters). Now have students determine the density of cattails (or trees) by dividing the number of cattails (or trees) by the area of the wetland. In the same manner, other sampling methods may be practiced using the poster.

Young students should trace a wetland area with their finger and count the number of cattails (or trees) in the wetland.

Activity

With students working in pairs or small groups, try the assessment techniques that seem most appropriate. Suggested grade levels are indicated with the activities presented in order of increasing skill level. Student math skills for grades K-2 are likely not sufficiently developed for these activities. Field guides should be used as needed to identify the plant species being examined.

Grades 3-12

A. COVER TYPE

Using the classification system at the bottom of any NWI map, describe the cover type or types for your wetland. Record the **Cover type** on the *Plant Community Summary* Student Page, and compare with previous years to determine if the cover type for your wetland has changed.

Grades 5-12

B. PLANT DENSITY

Density is the number of individuals per unit area. Using the data recorded on the *Plant Summary* in the column for **Survival**, add the number of live plants of all species within the sample area. Divide the total number of plants by the **Area Sampled** in square meters (m^2); this is the **Density** of **plants/m^2**. Record this on the *Plant Community Summary* Student Page. Separate densities may be calculated for trees, shrubs, vines, grasses, emergent plants, and submerged aquatic plants. How might this change over time? Why?

Density = total number of plants ÷ sample area

C. VEGETATIVE COVER

Add the data under **Plant Cover** on the *Individual Plants* Student Page to determine the total area covered by plants. What are the overall dimensions (length and width) of the area in which the plants are located?

The percent vegetative cover is equal to the total area covered by plants divided by the total area sampled. Record the **Percent Vegetative Cover** on the *Plant Community Summary* Student Page.

Percent vegetative cover =

total area covered by plants ÷ total area sampled

D. SPECIES RICHNESS

The number of different species of plants present in a community indicates richness. Identify and list the plant species found in the wetland, without damaging the wetland. How many total species were found in the wetland? Count them. This is a measure of **Species richness**. The greater the number of species, the richer the community. Record this information on the *Plant Community Summary* Student Page and keep the list of species for year to year comparisons. Are some species moving into or disappearing from the wetland?

E. SPECIES ABUNDANCE

Species abundance is the number of individuals per species. Count how many individuals of each species are present in the wetland or in representative samples of the wetland, such as quadrants or transects. Rank the species from most to least abundant, and graph the frequency distribution as a bar graph on the *Species Abundance Graph,* beginning with the most abundant species on the left side, then the second most abundant species, etc. Graph the twenty most abundant species of plants. Make a key for the graph.

The top one or two species are quite abundant and therefore dominant species. The next several species are considered moderately abundant. The rest are rarer, but contribute to species richness. Have the dominant species changed over time? Why or why not?

F. SPECIES FREQUENCY

Frequency is the proportion of sampling units in which a species occurs. It is a measure of the evenness of a species' distribution. Make a grid pattern to divide the wetland into quadrants; the quadrant size will depend on the size of the wetland. If the wetland is large, randomly select a number of quadrants to investigate. One method of making random selections is to assign a number to each quadrant. Write the numbers on slips of paper and place them in a container. After mixing them, draw a representative number of slips of paper; these will be the quadrants investigated in detail. If the wetland is small relative to the class size, all quadrants may be examined. If the vegetation is relatively short, an alternative method for selecting sample areas is to randomly toss hoops (approximately one meter diameter) into the wetland, then sample the plants within the hoop.

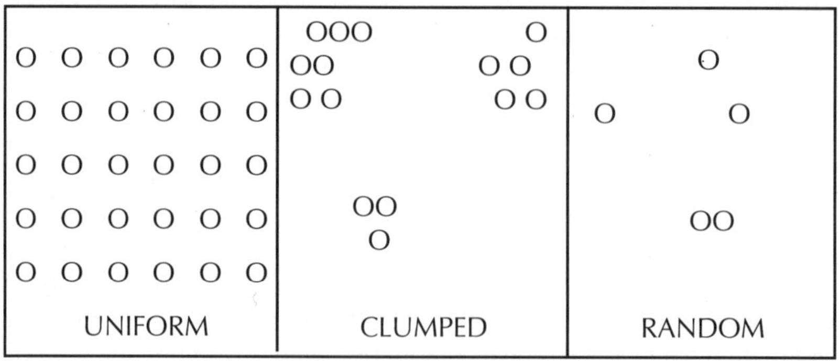

	OOO O	
O O O O O O	OO O O	O
O O O O O O	O O O O	O O
O O O O O O		
O O O O O O	OO	OO
	O	
O O O O O O		
UNIFORM	CLUMPED	RANDOM

Figure 10.2 Three typical distribution patterns are shown.

Now, using a list of the ten most abundant species from the previous activity SPECIES ABUNDANCE determine which of those ten most abundant species are present in the quadrants sampled. Record the frequency of each species first as a proportion (for example, Species #1, the most abundant species, was present in 6 of 12 quadrants), then as a percent (for example, Species #1 was present in 6/12 or 0.50 or 50% of the quadrants). List the overall ten most abundant **Species** (from Activity E) and their **Frequency As Percent** (as determined in this activity) on the *Plant Community Summary* Student Page. Why might species' frequencies change from year to year?

While sampling, notice how species are distributed throughout the wetland (see **Figure 10.2**). Are some species uniformly distributed? Are some species clumped together? Are some randomly distributed (not clumped or uniform)? Record the **Distribution Pattern** on the *Plant Community Summary* Student Page.

Grades 9-12
G. PRODUCTIVITY
Using the survival and biomass data recorded on the *Plant Summary* Student Page, multiply the **Average Biomass Per Plant** for each species times the number of **Live** plants of that species to determine the species' biomass in grams (g). Add together the biomass of all the species and divide by the **Area Sampled** in meters squared (m^2) to determine **Productivity** in g/m^2. Record the information on the *Plant Community Summary* Student Page. What might cause the productivity to change from year to year?

Species 1 biomass =
average biomass of Species 1 plants × number of live Species 1 plants

Productivity =
Species 1 biomass + Species 2 biomass + . . . ÷ Area Sampled

Wrap-Up

Store all the data collected in this activity to compare with future years. Our understanding of wetlands is evolving based on knowledge gained from such accumulated data.

What changes are occurring in the wetland plant community from season to season, and year to year? Describe these changes and hypothesize reasons for them. Are these changes acceptable? If not, what can be done to prevent these changes?

ASSESSMENT

Have students do the following:

- Draw or name the (three to ten) most abundant plant species in the wetland and list their characteristics.
- Orally or in writing describe the vegetative patterns discovered in the wetland.
- Is the wetland successful and healthy? What data support this evaluation?
- Compare data collected with previous data and give reasons for the changes that have occurred in the wetland over time.
- Hypothesize how the wetland plant community will change in the future, giving reasons for the changes.

EXTENSIONS

If this is the first year for the activity, have students create separate line graphs showing percent vegetative cover, productivity, plant density, species richness, and species frequency. Set up the graphs for a ten-year period, but only enter the data points for the first year. Five to ten species can be plotted on each graph if different colors or symbols are used for the data points.

If this is not the first year for the activity, have students add their data points to the graphs that have already been created.

RESOURCES

Caduto, M.J. 1990. *Pond and Brook.* University Press of New England, Hanover, NH.

Firehock, K., L. Graff, J.V. Middleton, K.D. Starinchak, and C. Williams. 1998. *Save Our Streams: Handbook for Wetlands Conservation and Sustainability.* Izaak Walton League of America, Gaithersburg, MD.

Kent, D.M. 1994. *Applied Wetlands Science and Technology.* CRC Press, Boca Raton, FL.

Project Learning Tree, Environmental Education Activity Guide. 1995. American Forest Foundation, Washington, DC. Especially: Trees in Trouble and How Big Is Your Tree?

Tiner, R.W. 1998. *In Search of Swampland: A Wetland Sourcebook and Field Guide.* Rutgers University Press, New Brunswick, NJ.

For free copies of the poster *Wetlands: Water, Wildlife, Plants and People!*, contact the U. S. Geological Survey, P. O. Box 25286, Denver Federal Center, Denver, CO 80225 or telephone 1-800-435-7627.

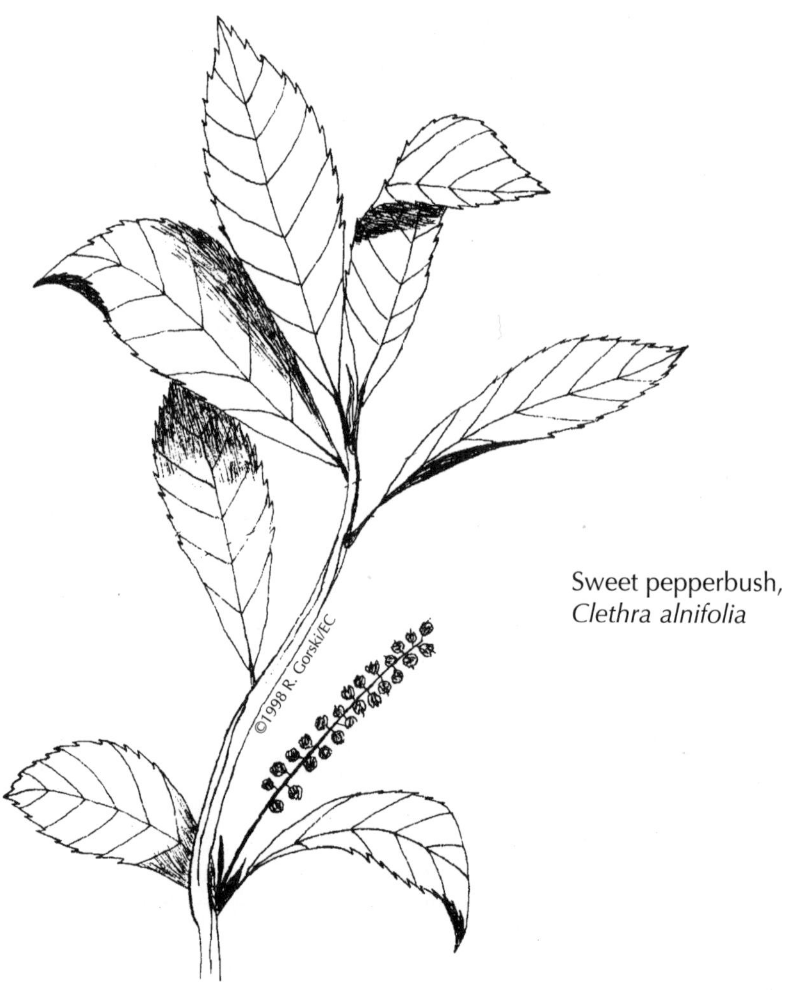

Sweet pepperbush, *Clethra alnifolia*

PLANT COMMUNITY SUMMARY

Date: _______________

A. Cover type: _________________________

B. Density: _______________ plants/m^2 _______________ trees/m^2

_______________ emergents/m^2 _______________ shrubs/m^2

_______________ submergents/m^2 _______________ vines or grasses/m^2

C. Percent vegetative cover: ______________

D. Species richness: _____________________

E. See *Species Abundance Graph.*

F. Species frequency:

Species	Frequency As Proportion	Frequency As Percent	Distribution Pattern
1.			
2.			
3.			
4.			
5.			
6.			
7.			
8.			
9.			
10.			

G. Productivity: __________________ g/m^2

SPECIES ABUNDANCE GRAPH

Date: ____________________

Number of Plants Per Species

40																				
38																				
36																				
34																				
32																				
30																				
28																				
26																				
24																				
22																				
20																				
18																				
16																				
14																				
12																				
10																				
8																				
6																				
4																				
2																				
0																				
	A	B	C	D	E	F	G	H	I	J	K	L	M	N	O	P	Q	R	S	T

Species

Soggy Soils

SUMMARY

Soil properties such as color and texture, soil moisture content, soil organic content, and development of soil horizons may be examined to assess hydric conditions.

OBJECTIVES

Students will analyze soil samples to determine their physical properties, organic content, and moisture levels. Soil horizons will be assessed for hydric properties. Changes in the wetland soil over time will be examined and evaluated for development of hydric properties.

MATERIALS

- Soil probe, soil shovel, or garden trowel
- Soil color charts
- Crayons
- Drying oven
- Balances
- Soil test kits for pH, nitrogen, phosphorus, potassium
- Copies of *Key to Soil Texture* (page 134)
- Copies of *Soggy Soils* Student Page

MAKING CONNECTIONS

Hydric conditions past and present can be determined with soil analysis. This can be important information, whether the soil in question is crop land, a construction area, a housing development, or a wetland. The techniques of soil analysis can be used for many purposes and situations.

BACKGROUND

Soils change slowly over time, but they do change. Usually soils develop layers, called horizons, made up of organic and mineral matter. In some of the wettest wetlands, a deep layer of organic material sixteen or more inches deep may have formed because plant matter is produced faster than it decomposes under anaerobic conditions. Organic soils are less dense than mineral soils, but have a greater capacity for holding water. They are acidic, easily penetrated by a soil borer or auger, and are called peat or muck. Peat is decayed organic matter in which the leaves, stems, and roots of plants are still identifiable when the material is rubbed between fingers. Peat is often used by gardeners as a planting material. Muck is organic matter in which very little is identifiable and the material becomes greasy when rubbed between fingers. Muck was an essential ingredient in the preparation of early cranberry bogs.

Mineral soils may also be hydric if they have an upper layer of organic matter eight or more inches deep; or if the mineral soil layer has orange, yellow, or red-brown mottles (specks) and the matrix (predominant soil material) has a chroma of two or less; or if the mineral soil layer has no mottles, but the matrix has a chroma of one or less.

Under the anaerobic conditions of permanent or semipermanent flooding, mineral soils become reduced through oxidation-reduction (redox) reactions. This results in gleying (when iron is chemically reduced) and is evidenced by a low chroma color. Chroma is an indication of color strength or purity; grayer colors have lower chroma values and are on the left side of standard color charts. Typically a chroma of two or less is an indication that the soil is hydric.

When mineral soils become alternately wet and dry, mottles form; these are specks of color within the larger gleyed soil matrix. Like the rust that forms on garden tools that are alternately wet and dry, the bright orange/red mottles are iron oxide, while the darker mottles are manganese oxide. Mottles will break apart when pinched between finger nails, but are relatively insoluble in water.

Another characteristic of hydric soils is the presence of oxidized rhizospheres. When oxygen leaks from the rootlets of hydrophytic plants, the iron in the soil is oxidized and bright orange/red strands appear within the gleyed soil matrix.

In a simplified wetland soil profile, the upper soil layer is an oxidized zone of organic matter. The next zone is formed by fluctuating ground water levels (sometimes wet, sometimes dry) and, therefore, contains mottles. The deeper zone is always wet, so the soil appears gleyed. Moving down a slope, these zones come closer to the surface. In a wetland the upper two zones may be very small.

If a mineral soil has sandy texture, then determining whether the soil is hydric becomes more difficult due to the rapid rate at which water will percolate downwards (high permeability). Evidence of saturation for an extended period must be found in the top six inches of sandy soil for it to be considered hydric. Characteristics that indicate that sandy soils are hydric are: a dark organic layer, muck, gleying, or dark organic streaking through a gray matrix.

PROCEDURE

Warm-Up

The day before the activity, collect a small soil sample from the schoolyard. Place the sample in a pint or quart jar with a lid (make sure the sample fills the jar no more than one third full). Add an equal amount of water to the jar and shake vigorously. Ask students to predict what will happen if the container is allowed to sit overnight undisturbed. Place the container in a prominent place in the classroom overnight. (Several small samples may be prepared and placed around the room to allow students to more easily see what has happened.)

Ask students to observe the container without disturbing it. What has caused the layering? (The particles have stratified according to particle size, with gravel in the bottom, sand next, then clay, and silt on the surface.) Observe the colors of the layers. Why are they different? (They formed from different parent materials.) When flooding occurs, soils that are washed downstream may stratify in a similar manner in quiet pools. Larger heavy particles like gravel will drop out first, while silt will stay suspended longer than other particles. In this manner soil particles are gradually moved from the eroding mountains to the oceans, slowly shaping our landscape in the process.

Activity

With students working in pairs or small groups, try the assessment techniques that seem most appropriate. Suggested grade levels are indicated, with the activities presented in order of increasing skill level. If chemical kits are used, students should follow safety precautions regarding the wearing of protective goggles and gloves.

Soil samples can be collected using a soil probe, soil shovel, or garden trowel. A soil probe extracts a deep narrow soil sample with minimal disruption of the landscape, but does not work well in dry soil or soil containing gravel. Samples collected with a soil probe can be laid out on a piece of plastic or cardboard for examination. Soil probes are available in many lengths. A soil shovel is narrower than a typical garden shovel, and can be used to dig a small pit; samples can be removed from the wall of the pit for student analysis. Garden shovels and/or garden trowels may also be used to extract soil samples, but obtaining a continuous sample will be more difficult. A continuous sample from the surface to a depth of eighteen inches (or as deep as can be reasonably achieved with students) should be examined for hydric properties, and the data that is collected compared with data collected from samples in previous years.

<u>Grades K-12</u>

A. SOIL COLOR

Using a soil color chart, determine the soil colors at two inch (5 cm) intervals along the soil sample, beginning with the surface of the sample. Does the soil sample show characteristics of hydric soils, such as gleying, mottling, oxidized rhizospheres, or rotten egg smell? If so, how far below the surface? Is this a change from previous years? Record soil **Color** and **Hydric Characteristics** on the *Soggy Soils* Student Page.

B. SOIL HORIZONS

Cut a strip of paper the length of the soil sample (18 inches or 45 cm). Have students use crayons to color the strip to match the horizons of the soil sample at each depth. Test colors on scrap paper before using them on the soil color strip. On the back of the strip mark the date the soil sample was collected and where it was collected. Has there been a change in the depth of the soil horizons since the previous year? Is the soil more or less hydric? Store the soil color strip for comparison in future years.

<u>Grades 3-12</u>

C. SOIL TEXTURE

Cut the soil sample into two-inch (5 cm) sections. Using the *Key to Soil Texture*, determine the soil texture of each two-inch section of the soil sample. (Follow the directions on the chart.) Is the soil sandy? Record the **Texture** on the *Soggy Soils* Student Page.

<u>*Grades 5-12*</u>

D. SOIL MOISTURE

Collect a duplicate soil sample, and again cut the sample into two-inch (5 cm) sections. Determine the mass of each sample on a balance and record it. Dry each section slowly in a warm (not hot!) oven. Again determine the mass of each section on a balance. Subtract the dry mass from the wet mass of each section; this is the amount of moisture present. Divide the amount of moisture by the wet mass of the section; this is the percent moisture for that section of the soil sample. Record the **Percent Moisture** on the *Soggy Soils* Student Page. Are some sections wetter than others? Why might this be the case?

moisture = wet mass – dry mass

percent moisture = moisture ÷ wet mass

E. SOIL pH

Using soil pH test kits, determine the acidity or alkalinity of each two-inch (5 cm) section of the soil sample. Since each test kit is slightly different, follow the directions that came with your test kit, but be sure it is one that is designed for soils!

Low numbers on the pH scale are acidic; high numbers are alkaline (basic). A pH of 7.0 is neutral. Soils with high levels of organic matter tend to be more acidic, but fewer plants are adapted to acid soil conditions. Does the pH change from the surface section to the bottom section? If so, how? Is your wetland soil acidic? Record the **pH** on the *Soggy Soils* Student Page.

F. SOIL NUTRIENTS

Fertilizers contain nitrogen, phosphorus, and potassium, which typically are the soil nutrients that limit plant growth when they are low or absent from the soil. Using soil test kits for nitrogen (N), phosphorus (P), and potassium (K), determine the level of nutrients available for wetland plants in each two-inch (5 cm) section of the soil sample. Follow the kit directions carefully. Are any of the nutrients low? Are any unusually high? If so, what might be the source of these excess nutrients? Record the **Nitrogen**, **Phosphorous**, and **Potassium** levels on the *Soggy Soils* Student Page.

Wrap-Up

If several soil samples were collected, were there differences between them? If so, how were they different?

Have changes occurred in the wetland soil over the years it has been tested? If so, how? Store the data collected in these activities for comparison with past and future years. Our understanding of wetlands is evolving based on knowledge gained from such accumulated data.

ASSESSMENT

Have students do the following:

- Draw, or describe orally or in writing, the differences in soil color, texture, and other characteristics from the surface of the sample to the bottom.

- Give possible reasons for the differences between the surface and the bottom sections.
- Are the differences caused by hydric conditions or by the normal weathering of soils? Explain your answer.
- Compared with soil sample data from previous years at the same location, have changes occurred? Why or why not?

EXTENSIONS

Compare and contrast soil samples taken from the edge of the wetland, up slope from the wetland, and down slope from the wetland. Use whatever tests are appropriate. Give reasons for the similarities and differences among samples. If necessary, locate references on soils for more background on the topic.

Invite a representative of the county Soil Conservation Service to bring local soil maps to school. Have that person explain how to read the maps, tell what the symbols represent, and describe how the maps are used.

RESOURCES

Firehock, K., L. Graff, J.V. Middleton, K.D. Starinchak, and C. Williams. 1998. *Save Our Streams: Handbook for Wetlands Conservation and Sustainability.* Izaak Walton League of America, Gaithersburg, MD.

Kent, D.M. 1994. *Applied Wetlands Science and Technology.* CRC Press, Boca Raton, FL.

Mitsch, W.J. and J.G. Gosselink. 1993. *Wetlands.* Van Nostrand Reinhold, New York, NY.

National Research Council. 1995. *Wetlands: Characteristics and Boundaries.* National Academy Press, Washington, DC.

Tiner, R.W. 1998. *In Search of Swampland: A Wetland Sourcebook and Field Guide.* Rutgers University Press, New Brunswick, NJ.

Soil supplies are available from:

Forestry Suppliers, Inc.; P.O. Box 8397, Jackson, MS 39284-8397; telephone 1-800-647-5368 or fax 1-800-543-4203 for a catalog.

LaMotte Co.; P.O. Box 329, Chestertown, MD 21620; telephone 1-800-344-3100 for a catalog.

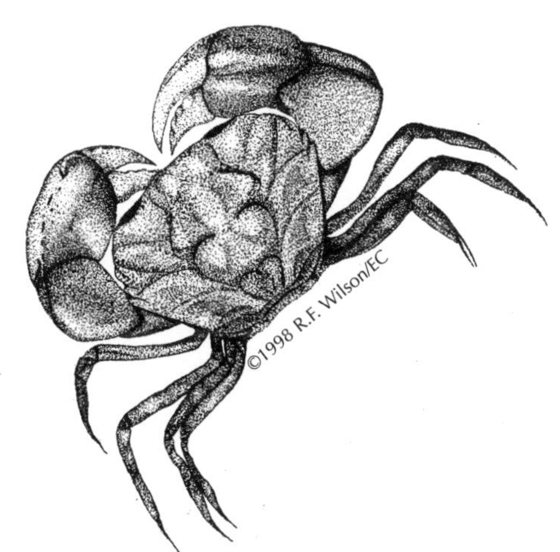

White-fingered mud crab, *Rhithropanopeus harrisii*

SOGGY SOILS

Sample site: ____________________ Date: ______________

Depth	Color	Hydric Characteristics			Texture	Percent Moisture	pH	Nitrogen (N)	Phosphorous (P)	Potassium (K)
		Gleyed	Mottled	Other						
0-2 in. (0-5 cm)										
2-4 in. (5-10 cm)										
4-6 in. (10-15 cm)										
6-8 in. (15-20 cm)										
8-10 in. (20-25 cm)										
10-12 in. (25-30 cm)										
12-14 in. (30-35 cm)										
14-16 in. (35-40 cm)										
16-18 in. (40-45 cm)										

Healthy Water

SUMMARY

The parameters of water quality will be measured and monitored for change over time. Differences in the quality of water entering, water within, and water leaving the wetland may be monitored and compared. Variables that may be measured include temperature, pH, turbidity, total solids, dissolved oxygen, biochemical oxygen demand, nutrient levels, and pollutant levels.

OBJECTIVES

Water quality parameters will be measured and analyzed for patterns of change. Potential causes of water quality change will be evaluated for positive and negative effects on the wetland habitat.

MATERIALS

- Thermometer
- Secchi disc
- Sample containers
- Balance
- pH test paper, pH test kit, or pH meter
- Test kits for nutrients such as nitrate and phosphate
- Dissolved oxygen test kit
- Additional dissolved oxygen sample bottles
- Test kits for pollutants
- Copies of *Water Quality* Student Page

MAKING CONNECTIONS

All life depends on water for survival. If the water is of poor quality or is contaminated, the quality of life for plants, animals, and humans is diminished. Measurement and analysis of water quality are important for assessing sources of drinking water, determining treatment levels for wastewater, and evaluating habitat quality for plants and wildlife.

Throughout our lives we are faced with water quality issues that require our understanding and evaluation. For example, the presence of any mercury in our drinking water would indicate a risk to health; the presence of some fluoride is considered an asset to health. An appropriate amount of chlorine in our drinking water kills unwanted bacteria, while excessive amounts of chlorine can cause problems. Measuring the quality of water in a wetland can help us determine what types of contaminants are present, and which ones are present at harmful levels.

BACKGROUND

The quantity and quality of water present have been determining factors in the location of human settlements, plant communities, and animal colonies throughout history. Water supplies that are healthy for one group, however, may not be suitable for others.

PROCEDURE

Warm-Up

Before class, prepare three containers of water. Add a small amount of sugar to the first one, some salt to the second one, and yellow food coloring to the last. Use a quantity of sugar and salt that can be tasted, but not seen when stirred into solution. The food coloring should only be slightly visible when mixed.

Tell students that they are evaluating three water sources to determine whether they are suitable for drinking water. Supply each student with a paper cup and a small sample of water from one of the three containers. Have students write or state their observations of the water sample, especially of color, odor, and taste. If possible, have each student evaluate all three water sources (paper cups should be rinsed between samples).

Which water sources are preferred by the group? Why? Could something be done to improve the water quality of all or some of the water sources? If so, what? (Accept all reasonable suggestions.)

Activity

With students working in pairs or small groups, try the water assessment techniques that seem most appropriate. Suggested grade levels are indicated, with the activities presented in order of increasing skill level. If chemical kits are used, students should follow safety precautions on the wearing of protective goggles and gloves.

Many of the tests involve color comparisons. Be aware that students who are colorblind may not be able to discern these color differences.

Grades K-12

A. WATER TEMPERATURE

The first time water temperatures are examined, many locations should be tested to determine temperature patterns within the wetland. Suggested locations to test include the water's edge, surface water in the middle of the wetland, surface water over the deepest part of the wetland, and bottom water in the deepest part of the wetland. Also test the inlet and outlet of the wetland, if these locations exist.

If more than one thermometer is being used, make sure that they agree. Even better, calibrate the thermometers. If the range of the thermometers includes the freezing (32°F or 0°C) and boiling (212°F or 100°C) points of water, place the thermometers in ice water until their readings stabilize. Any thermometer that does not accurately reflect the temperature should be set aside and probably discarded. Then warm the water bath slowly on a hotplate until the boiling point is achieved. Observe the temperature readings of the thermometers, carefully using tongs, gloves, towels, or something to protect hands from burns. Again, any thermometer that does not accurately reflect the temperature should probably be discarded, unless it can be adjusted appropriately. (If it is adjustable, directions would have been supplied with the thermometer when purchased.) The degree of precision necessary depends on the ability of students to read a thermometer accurately. If students are able to read a thermometer only to the closest degree, then a thermometer need only be accurate to within 0.5° of the actual temperature.

To test surface water temperatures, simply place the sensitive bulb end of the thermometer in the water. Hold it there until the temperature reading stabilizes, then report and record the **water temperature**. Do not hold the thermometer out of the water to read the temperature, unless it can be done rapidly; the temperature will change to the air temperature. Also, do not handle the bulb end of the thermometer because this will result in a reading of body temperature instead of water temperature.

To test bottom-water temperature, a water sampling device may be purchased and used according to the manufacturer's directions. If the water is shallower than an adult's arm length, any container that holds 250 ml of water or more, and is easily opened and closed, will do. Close the empty container. Reach down to the bottom waters. Open the container, allow it to fill with water, then close the container. Bring the container of water to the surface, open it, and insert the thermometer. Once the temperature reading has stabilized, record the **water temperature**.

Now hold the thermometer in the air, being careful not to hold it by the sensitive bulb end. When the temperature reading stabilizes, report and record the **air temperature**.

All temperatures may be recorded on the *Water Quality* Student Page, along with the date and time of day. Does the temperature differ between the surface and bottom of the wetland? What might cause this? Is the temperature at the edge different from the temperature in the middle of the wetland? What might cause this? Are the temperatures at the inlet and outlet of the wetland different? Why? (Answers will depend on the time of year, time of day, size of wetland, water source, and many other factors that characterize each wetland.)

Select locations to be tested on a regular basis (weekly or monthly). Temperature data may be plotted on a graph using colored markers to represent different locations. How does the water temperature at each location change through time? Why? What patterns of change can be detected? What is producing these patterns? Are these same patterns of change likely to occur at other wetlands? Why or why not?

B. TURBIDITY

Turbidity is a measure of water clarity or of the distance that light can penetrate the water. If the bottom of the wetland is always visible, turbidity is very low. If the bottom of the wetland is not always visible, turbidity is at least occasionally high. Plants need light to photosynthesize. Which types of plants would be affected by turbidity?

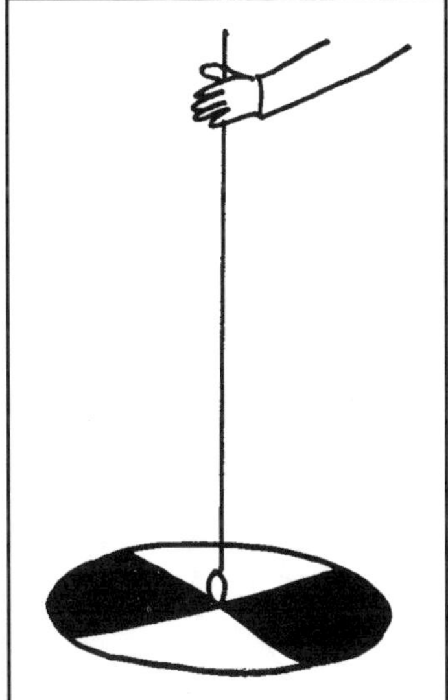

Figure 10.3 A secchi disc can be made from a paint can lid.

Turbidity can be measured either with a secchi disc that is purchased or constructed or with a water testing kit that uses chemicals. To construct a secchi disc, obtain the lid of a one-gallon paint can. Divide the lid into four pie-shaped quarters; paint two opposite quarters black and two opposite quarters white (see **Figure 10.3**). Drill a small hole in the center of the lid and attach an eye bolt and nut. Securely tie a line to the loop end of the eye bolt on the painted side of the secchi disc. Attach enough washers (using the nut on the underside of the secchi disc) to help the disc sink in water.

Lower the secchi disc into the water by the attached line until it is no longer visible. Now raise the disc until the black and white quarters can just barely be seen. Grab the line with two fingers where it enters the water; hold that mark as the disc is lifted out of the water. Measure the distance from the two fingers holding the line to the disc. This is the depth of light penetration or **turbidity**. Record this measurement, the date, and the time of day on the *Water Quality* Student Page. Could the measurement differ according to the time of day? Why? Compare measurements made at different times on the same day to test your idea.

Compare the turbidity measurement with others taken in the past. Does the level of turbidity change weekly or with the seasons? Why? Is the wetland more turbid after storms? What might cause this? Graph the turbidity level and analyze the graph for patterns of change through time.

Grades 5-12

C. pH

Some wetlands, such as bogs, are naturally acidic because of the plants that grow there. Some wetlands are acidic due to the soils and rocks present in the area. Stormwater runoff and precipitation may pick up chemicals on the ground and in the air, however, that contribute to wetland acidity in ways that are not natural or healthy for the organisms living there.

To determine the acidity of wetland water and to monitor it for change, pH paper, a pH chemical test kit, or a pH meter is needed. The pH scale ranges from 0 (highly acidic), to 7 (neutral), and to 14 (highly alkaline or basic). Most living organisms prefer water that is neutral; fewer species of plants and animals are adapted to living in more acidic or alkaline waters.

The easiest and safest measure to use is pH paper or litmus paper; it is recommended for use with younger students but is not very precise. A short strip of paper is removed from the tape-like dispenser and dipped into the water being tested. The color of the wet paper is then compared to the color scale on the dispenser to determine the pH of the water.

A pH chemical test kit uses a variety of chemical indicators. More caution is required, but results are more accurate. Each kit contains clearly printed directions that are easy to follow. Replacement chemicals are available in bulk quantities for resupplying the kits.

Meters are the most accurate devices for measuring pH, but they require periodic calibration and are expensive. They are available in small box-size units such as are used by professionals, and as pens the size of small flashlights. This may be an appropriate choice for testing by students in the upper grade levels. Each meter comes with appropriate directions for use and calibration. During calibration, the probe is placed in a buffer solution of known pH and the scale on the meter is adjusted to agree with the pH of the buffer solution.

Test pH at the same locations where temperature is tested. Record all **pH** measurements, the date and time of day on the *Water Quality* Student Page. Are there differences in the measurements taken at different locations in the wetland? Why or why not? What could be the cause or source of increased acidity, if it exists?

Record the data on a graph using different colors for each location. Provide a key to the colors used. How has pH changed through time? Are there patterns for the change or is it unidirectional? What might be the cause? How might this affect the plants and animals that utilize the wetland?

D. TOTAL SOLIDS

Whether suspended or dissolved, solids in the water can cause it to appear turbid and can prevent light penetration. To measure the amount of total solids, first label several sample containers without lids: A, B, C, etc. Use a balance to measure the mass of each container; record the **mass of empty container**. Collect equal-sized quantities of water (100 ml) at each sample location. Record where each sample was collected along with the letter of the container, but do not write anything on the container, since this will change its mass. Place the containers of water in an area where they will not be disturbed, and allow the water to evaporate leaving behind the dissolved and suspended solids.

When the sample containers are completely dry, determine the **mass of container and dry solids**. Subtract the mass of the empty container from the mass of the container and the dry solids. The difference is the **mass of total solids** (both dissolved and suspended) in each sample. Record the measurements on the *Water Quality* Student Page.

Mass of total solids =
mass of container and dry solids – mass of container

Which sample is the heaviest, and where was it taken? Which sample location has the least amount of solids? Compare with previous measurements. Are there differences in the quantities of solids in the water? If so, what might be the cause?

Construct a line graph showing the total solids determined at each sample location and the dates of the collections. If samples were collected at several locations, use a different color marker for each location and make a key. Are there patterns of change over time? What might be the causes?

E. NUTRIENTS

Plants need nutrients such as nitrate and phosphate for normal healthy growth. If essential nutrients are not available in sufficient quantity, plant growth is limited. Usually this is not a problem in wetlands, since wetlands filter and trap excess nutrients that run off lawns, gardens, farm fields, golf courses, and livestock areas. Some wetlands have even been created as specialized sewage treatment systems. While excess nutrients can be a problem in some habitats, excess nutrients in a wetland may indicate that the wetland is performing as it was designed by removing excess nutrients before they enter a waterway. To determine what is happening in your wetland, measure nutrient levels at the place where water enters the wetland, within the wetland, and at the place where water leaves the wetland.

Nutrient levels can be measured using test kits available from several suppliers. The test kits contain pre-measured dry chemicals in packets, liquid chemicals in plastic dispensers, color comparison charts or plastic devices, and simple step-by-step directions.

Nitrate and phosphate are present in fertilizer and animal waste, so they should be measured. Many other nutrient levels can also be measured, such as potassium and nitrite.

Using clean sample containers, collect water samples where water enters the wetland, within the wetland, and where water leaves the wetland. Using the test kits, determine the amount of each nutrient present. A natural reading for nitrate is 0.00 to 0.65 ppm (parts per million); a natural reading for phosphate is 0.00 to 0.08 ppm. Did test results indicate higher than normal amounts of these chemicals? If so, at which locations? What might be happening within the wetland? What are the possible sources of these nutrients? Record the test results for **Nitrate** and **Phosphate**, date, and time of day on the *Water Quality* Student Page.

Compare current test results with those collected previously. Have they changed? If so, why? Graph the test results to show how the nutrient levels change through time. If more than one site is being tested, use different color markers for each, and provide a key for the colors used. Are patterns of change evident when the graph is analyzed? What may be the cause of the change? Explain.

F. DISSOLVED OXYGEN

Animals need oxygen to live, and aquatic animals need oxygen dissolved in their water. Wind, water movement, and photosynthesis are sources of dissolved oxygen, which is used by aquatic animals, plants, and bacteria for respiration. Factors such as temperature, wind, and weather influence the amount of dissolved oxygen in the water. Cold water is able to hold more dissolved oxygen than hot water, and fresh water can hold more dissolved oxygen than salt water.

When excess nutrients are present, an algal bloom may occur. During the night, algae respire, using the dissolved oxygen produced during the day. When the algae die, decomposing bacteria use large amounts of dissolved oxygen during respiration as they feed on the dead algae. Algal blooms, therefore, usually deplete dissolved oxygen supplies.

Because of these many interactions, dissolved oxygen levels indicate water quality. Most aquatic organisms need at least 5 ppm (parts per million) dissolved oxygen; 3 to 4 ppm results in stress; 1 to 2 ppm is functionally anoxic, and 0 ppm is anoxic (without oxygen).

DO (dissolved oxygen) test kits and DO meters are available from several supply companies. The kits provide pre-measured dry chemicals in packets, liquid chemicals in plastic dispensing bottles, sample bottles, and clear directions. DO meters have probes that must be standardized and calibrated. The meters require more care than the test kits, but measurements are more precise. For best results, kits or probes must be taken into the field and used where and when the samples are collected. Clean plastic laundry detergent jugs make handy waste containers that readily accept used chemicals and used chemical packets, which can then be transported back to the classroom where chemicals can receive proper disposal.

Measure the dissolved oxygen levels at the same locations that temperature is measured. Record the **dissolved oxygen** levels, the date, and time of day on the *Water Quality* Student Page. Does the wetland have healthy levels of dissolved oxygen at all locations tested? Why or why not? How will this affect the creatures in the wetland?

Test the dissolved oxygen level at one location several times during a single day. Chart the change on a graph of DO level and time of day. How does the dissolved oxygen level change throughout the day? What might cause this? How might this affect the creatures in the wetland?

Compare current DO measurements with past levels by compiling the data into a graph of DO versus week or month. (This is more meaningful if all measurements are taken at the same time of day.) Are patterns of change evident? Are there differences between seasons? What might be the cause? How might wetland creatures adapt to these changes?

G. BIOCHEMICAL OXYGEN DEMAND
Biochemical oxygen demand (BOD) is caused by microbes consuming oxygen as they break down organic matter to obtain energy through the process of respiration. BOD is a measure of the amount of decomposition occurring in a wetland.

A DO test kit and extra DO sample bottles are needed for this activity. Collect two water samples in DO bottles at each location to be tested. Test the DO level in sample A immediately; record the **DO of sample A** on the *Water Quality* Student Page. Wrap sample B in foil or other light blocking material, then place it in a dark cabinet at room temperature for five days. (If an incubator is available, store the sample in the dark at 20°C for five days.)

After five days, remove the light-blocking material and test the DO level; record the **DO of sample B** on the *Water Quality* Student Page. Subtract the DO level measured in sample B from the DO level measured in sample A; this is the BOD. The units of measure may be parts per million (ppm) or milligrams per liter (mg/L); these are equivalent units. Record the **BOD**, the date, and the time of day that samples were collected on the *Water Quality* Student Page.

BOD = DO of sample A – DO of sample B

1 ppm = 1 mg/L

Take duplicate water samples at one location several times during a single day. Test the BOD levels. Chart the change on a graph of BOD level and time of day. How does the BOD level change throughout the day? What might be causing this? What is likely to be happening at night to the DO levels? How might this affect the creatures in the wetland?

Compare current BOD measurements with past levels by compiling the data into a graph of BOD and week or month. (This is more meaningful if all measurements are taken at the same time of day.) Are patterns of change evident? Are there differences between seasons? If so, what might be the cause? How might wetland creatures adapt to these changes?

Grades 9-12

H. POLLUTANTS

Many types of pollution may be present in a wetland, especially if the water supply is stormwater runoff from a parking lot or street. List the land uses of the area surrounding the wetland; then determine what pollutants might be present. Select chemical test kits that will help measure levels of suspected pollutants. Carefully follow the directions of each test kit. Sample water where it enters the wetland, in the wetland, and as it leaves the wetland.

Did test results indicate high amounts of the pollutant tested? If so, at which locations? What might be happening to these pollutants within the wetland? What are the possible sources of these pollutants? Record the test results, date, and time of day on the *Water Quality* Student Page.

Compare current test results with those collected previously. Have they changed? If so, why? Graph the test results to show how the pollutant levels change through time. If more than one location is being tested, use different color markers for each location and provide a key. Are patterns of change evident when the graph is analyzed? What may be the cause of the change? Explain.

Wrap-Up

Does water quality improve between entering the wetland and exiting it? If one goal for the wetland was to improve water quality, is it performing as planned? If not, why not? Do the measured changes in water quality have a positive or negative effect on the wetland? Do the changes in water quality correlate with changes in wetland vegetation, soils, or hydrology? If so, how?

Is water quality in the wetland improving or declining through time? Our understanding of wetlands is evolving based on knowledge gained from accumulated data, so it is important to keep summaries of all wetland data in a single location, such as a large binder or file drawer.

ASSESSMENT

Have students respond orally or in writing.

- Describe how water temperature, dissolved oxygen, and biochemical oxygen demand change throughout the day. What causes these changes?
- Of the water quality factors that were tested, which ones improved between water entering and exiting the wetland? Was this planned? What is causing the improvement?
- Among the water quality factors tested, which ones are improving through time? Explain why this is happening.
- Are any aspects of water quality declining? If so, hypothesize a cause and a solution.

EXTENSION

Investigate the effects of increased temperature, turbidity, acidity, and dissolved oxygen on the animals utilizing the wetland. Determine the effects of a decline in these factors on the animals.

Nitrate and phosphate levels, total solids, and biochemical oxygen demand are used to characterize wastewater treatment systems. The differences between levels at inflow and outflow indicate the effectiveness of the system. Investigate the use of natural systems such as wetlands in wastewater treatment. Compare the wetland being monitored with a natural wastewater treatment system.

One function of a wetland is to remove pollutants from the water. How are specific pollutants removed? Are they strained, filtered, absorbed, adsorbed, or removed by some other means?

RESOURCES

Caduto, M.J. 1990. *Pond and Brook.* University Press of New England, Hanover, NH.

Center of Occupational Research and Development (CORD). 1991. *Applications in Biology/Chemistry: Water.* CORD Communications, Waco, TX.

Reed, S.C., E.J. Middlebrooks, R.W. Crites. 1988. *Natural Systems for Waste Management and Treatment.* McGraw-Hill Book Company, New York, NY.

Kesselheim, A. S. and B. E. Slattery. 1995. *WOW! The Wonders of Wetlands.* Environmental Concern Inc. St. Michaels, MD. Especially: *Water We Have Here?* and *Nutrients: Nutrition or Nuisance?*

Water testing equipment and supplies are available from:

Carolina Biological Supply, 2700 York Road, Burlington, NC 27215; telephone 1-800-334-5551.

LaMotte Company, P. O. Box 329, Chestertown, MD 21620; telehone 1-800-344-3100.

Hach Company, P. O. Box 389, Loveland, CO 80539-0389; telephone 1-800-227-4224.

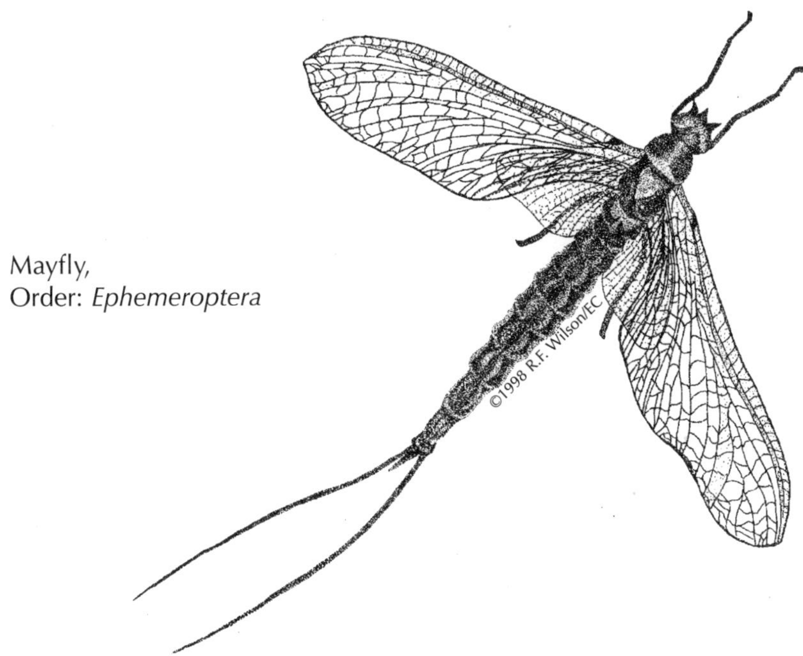

Mayfly,
Order: *Ephemeroptera*

WATER QUALITY Date: ______________ Time: ____________

Parameter	Site 1	Site 2	Site 3	Site 4
Air temperature (°C or °F)				
Water temperature (°C or °F)				
Turbidity (cm or in)				
pH				
Mass of empty container (g)				
Mass of container & dry solids (g)				
Total solids (g/100 ml)				
Nitrate (ppm or mg/L)				
Phosphate (ppm or mg/L)				
Dissolved oxygen (ppm or mg/L)				
DO sample A (ppm or mg/L)				
DO sample B (ppm or mg/L)				
BOD (ppm or mg/L)				

Marvelous Hydrology

SUMMARY

Water inflow to the wetland from natural water sources will be assessed over time to ensure that the wetland hydrology is functioning as planned. This activity is not appropriate if the wetland is supplied by an artificial water source.

OBJECTIVES

Students will measure the water inputs to the planned wetland, calculate water volumes, compare water volumes with those predicted to be available to the wetland, and evaluate whether sufficient water is available to the wetland for it to function as planned.

MATERIALS

- Clear box or jar
- Packing peanuts
- Meter stick or yard stick
- Fence post
- Permanent marker
- Copies of *Surface Water Depths* Student Page
- Rain gauge
- Local precipitation data
- Copies of *Precipitation* Student Page
- Four-inch diameter round plastic pipe and two-inch diameter wooden dowel of the same length (both should be taller than the highest water level)
- Screen to cover the end of pipe and duct tape
- Cork, ground into particles
- Coffee can
- Copies of *Storm Events* Student Page
- Post hole digger or six-inch auger
- Pea-size gravel to fill the hole two inches deep
- 24-inch length of four-inch diameter plastic pipe (e.g. schedule 40 PVC pipe) with a plastic cap
- Copies of *Groundwater Levels* Student Page

MAKING CONNECTIONS

Once a wetland has been constructed, it still must be evaluated and reevaluated to ensure that it is functioning as planned. If not, the investment of time, energy, and money that created the wetland may be lost—just as with any other investment. The initial investment is usually the largest, but small investments of effort and even money may be needed to keep whatever was created (a wetland, a house, a business, etc.) functioning as planned.

BACKGROUND

A planned wetland, whether it is created, restored, or enhanced, is designed to meet specific goals. If the amount of water actually available for a wetland differs substantially from what was predicted to be available, the wetland goals may not be met.

If insufficient water is available, then hydric conditions will not be present and wetland plants may not survive. Without wetland habitat, wetland creatures will not colonize the area. Without thriving wetland plants, other wetland functions such as pollution control, sediment retention, and flood control may not occur as planned.

If water is available far in excess of what was planned, the wetland may become larger or erosion may occur at the outflow if a water control device is not in place. In any case, a decision will need to be made whether to allow the expansion of the wetland, or to install or upgrade a water control device.

PROCEDURE

Warm-Up

Volume is a difficult concept. Have younger students predict how many marshmallows (or sugar cubes or beans or marbles) a particular container will hold. Record their predictions on the board. Add the marshmallows to the container one at a time as the students count. Stop when the container is full. Explain that volume can be measured (as in "the container is x marshmallows in size") or calculated if the length, width, and depth are known.

Volume = length × width × depth

If a miniature marshmallow is one centimeter wide, one centimeter long, and one centimeter deep in size, what is the volume of the miniature marshmallow? (1 cm × 1 cm × 1 cm = 1 cm^3)

How many miniature marshmallows will a box hold, if the box is ten centimeters long, eight centimeters wide, and four centimeters deep? (10 cm × 8 cm × 4 cm = 320 cm^3; 320 cm^3 ÷ 1 cm^3 = 320 miniature marshmallows.)

Activity

Select activities appropriate for your planned wetland. Suggested grade levels are indicated, with activities presented in order of increasing skill level.

Grades K-12

A. SURFACE WATER DEPTH, PART I

On a regular basis (once a week would be excellent!) record the water depth in the deepest area of the wetland. This can be accomplished by wading into the water with a meter stick (if the water is shallow) and measuring the depth. Even better, install a post marked with water depth measurements in the deepest part of the wetland. The post can be observed from outside the wetland and the water depth determined from a distance, even during flooding. Measurement marks on the post should begin with zero where the post goes into the ground surface, and increase up the post high enough to be above any expected flood levels. Marks should be readily visible at a distance. If measurements are in feet, marking tenths of a foot on the post (instead of inches) will facilitate calculations and graphing.

A marked post will allow easy determination of the maximum depth (if it is in the deepest area of the wetland), but it does not indicate the average depth. To determine the average depth (for water volume calculations), the water depth at the edge of the wetland must also be measured, then the two depths averaged. For example, if the marked post shows the greatest water depth to be 1.5 feet and the edge depth is 0.5 feet, then the average water depth is 1.0 feet (1.5 feet + 0.5 feet ÷ 2 = 1.0). Keep in mind that this is an approximation of average depth because of the limited number of depth measurements.

What volume of water is currently in the wetland? (Measure and calculate length times width times average depth.) Is this more or less than the planned volume of water calculated in #4 of the *Water Budget*?

Is the water depth more or less than the designed pool level in the *Grading Plan*? If the water depth is not what was planned, why is it different? How does the water depth change through time?

B. PRECIPITATION, PART I

Mount a rain gauge at the top of a post in an open area near the wetland. Be sure there is nothing above or beside the gauge to prevent it from collecting rain. Whenever precipitation occurs, check the rain gauge to determine the amount. Record the date and the amount, then empty the rain gauge. Check the rain gauge on Monday mornings for weekend precipitation.

Total the precipitation that occurs each month and compare with the average monthly precipitation listed in the local county soil survey publication prepared by the Natural Resource Conservation Service. The same information is available on the internet (see Resources). Compare the long term average monthly precipitation with the actual averages just calculated. Are they similar? If not, have any unusual weather events occurred?

Convert your measured average monthly precipitation to biweekly precipitation (divide monthly precipitation by two) or calculate actual average biweekly precipitation by adding together weeks 1 and 2, weeks 3 and 4, etc. How does the actual biweekly average precipitation compare with that predicted in #1 of the *Water Budget*? Save all data for comparison in future years.

Grades 5-12

C. SURFACE WATER DEPTH, PART II

Water depth may be tracked on a line graph each week of the season for the entire school year, then compared with previous years. A graph for *Surface Water Depths* is provided. Which seasons are wet? Which are dry? Is there ever insufficient water for the wetland to function as planned? Why? What other patterns are perceived?

D. PRECIPITATION, PART II

Precipitation may be tracked each season for the entire school year on a weekly or biweekly basis using a bar graph. A graph for *Precipitation* is provided. How does precipitation affect surface water depth in the wetland? Are there patterns of precipitation? Have precipitation patterns changed from previous years?

Grades 9-12

E. STORM EVENTS

Construct a crest height gauge to record the highest water level with each storm event. (Adapted from *Save Our Streams Handbook for Wetlands Conservation and Sustainability,* 1998.)

First, obtain a four-inch diameter plastic pipe and a two-inch diameter wooden dowel (the same length as the pipe), both of which will be taller than the highest water level when installed in the wetland. Cut a notch one half inch wide and four inches high in the bottom of the pipe so that water can enter the pipe even if sedimentation occurs around the base of the pipe. Cover the bottom of the pipe and the notch with a screen, such as 1/4 inch hardware cloth. Attach the screen with duct tape or wire, being careful not to cover the notch with tape. Mark the outside of the pipe with a permanent marker in tenths of a foot. Also mark the wooden dowel with the permanent marker in tenths of a foot. (Marking in tenths of a foot allows for easier calculations than if inches were used, and is an acceptable practice.)

Next, install a fence post in the wetland and attach the pipe with clamps or wire. The bottom of the pipe should just touch the ground surface. Be sure the notch is not against the fence post, since this is how water will enter the pipe.

Now sprinkle ground pieces of cork inside the plastic pipe. Rinse the inside of the pipe with a little water to remove any cork clinging to the sides. Place the wooden dowel inside the pipe and cover the top of the pipe with a coffee can.

When a storm occurs, the water level in the wetland will rise, lifting the cork inside the pipe. After the storm has passed, the level of the dried cork on the dowel can be measured to determine the highest water level in the wetland. After measuring, rinse the cork off the dowel and the inside wall of the pipe. Place the dowel back inside the pipe and the coffee can back on top of the pipe.

After each storm event, record the highest water level as measured with the crest height gauge (or the surface water depth of the wetland as in Activities A and C), and the amount of precipitation recorded in the rain gauge (as in Activities B and D). Subtract the surface water level before the storm from the highest water level produced by the storm; this is the change in water depth. Plot the change in water depth versus the amount of precipitation (as recorded in the rain gauge) on the graph of *Storm Events*. As more storm events are plotted on the graph, the scattered points will begin to form a line. With a pencil, draw one straight line that best fits the plotted points.

Have students answer the following questions:

a. At what surface water depth (or pool level) is the planned wetland considered full? (Check the *Grading Plan*.)
b. How much precipitation is needed for the wetland to fill with water if the wetland is empty? You may need to extrapolate (extend the line) on the graph of *Storm Events* to predict this.
c. What is the average biweekly precipitation calculated in #1 of the *Water Budget*?
d. Does the average biweekly precipitation alone supply the needed quantity of water for the planned wetland? If not, what are the other sources?

F. MONITORING WELL

If necessary, obtain a permit first; then install one or several monitoring wells to measure groundwater levels. (See **Figure 2.7** and pages 44-45. Also described in *Save Our Streams Handbook for Wetlands Conservation and Sustainability,* 1998)

Up slope of the wetland edge, use an auger or posthole digger to dig a narrow, twenty-inch deep hole. Save the removed soil. Add two inches of pea-size gravel to the hole; then measure to be sure that the distance from the top of the gravel to the top of the hole is eighteen inches.

Insert a 30-inch length of two-inch diameter, perforated, schedule 40 PVC pipe into the hole. Fill the area around the pipe to within two inches of the ground surface with gravel of 1/4-inch to 3/8-inch size. On top of this, firmly pack soil around the pipe to a height of four to six inches above the ground surface. Slope the edges to resemble an ant hill, making sure eight to ten inches of pipe extends above this packed mound. Cover the top of the pipe with a can. (If the pipe is fitted with a plastic cap instead of a can, drill a hole in the pipe just below the cap so air can escape or enter as the water level changes.)

On a regular basis (weekly or monthly), measure the water depth in the well. First, remove the coffee can or cap from the well. Using a flashlight, measure the distance from the top of the well to the water level. The distance from the top of the pipe to the water surface minus the distance from the top of the pipe to the ground surface outside the pipe, equals the depth at which water exists below the ground surface. Chart the groundwater level on the graph of *Groundwater Levels* with the distance from the ground surface to the groundwater on the x-axis and week (or month) on the y-axis.

How does the water level fluctuate throughout the year? Add the data on surface water depth (Activities A and C) to your graph, using a different color line (or add the ground water depth to the graph of *Surface Water Depths*). Are there correlations between the water level in the well and the surface water depth in the wetland? Is groundwater alone supplying sufficient water for the wetland to function as planned? If not, why not? Is there new construction in the area that may have affected the groundwater supply? Compare with previous years. Are there seasonal or annual patterns? Save data for future use.

Wrap-Up

Is there sufficient water available from natural sources for the wetland to function as planned during the growing season? If not, why not? What changes, if any, can be made?

Save all data in a binder or file drawer. Our understanding of wetlands is evolving because of knowledge gained from such accumulated data.

ASSESSMENT

Examine student papers.

- Are student calculations accurate?
- Are graphs accurately completed?
- Is reasoning logical?
- Are solutions creative?

EXTENSIONS

If this is not the first year for this activity, compare data with that of previous years. Are there annual patterns of surface water depth, precipitation, and groundwater levels? Is there a pattern to the size and frequency of storm events? When is the planned wetland wet and for how long?

RESOURCES

Firehock, K., L. Graff, J.V. Middleton, K.D. Starinchak, and C. Williams. 1998. *Save Our Streams: Handbook for Wetlands Conservation and Sustainability.* Izaak Walton League of America, Gaithersburg, MD.

Kent, D.M. 1994. *Applied Wetlands Science and Technology.* CRC Press, Boca Raton, FL.

Mitsch, W. J. and J.G Gosselink. 1993. *Wetlands.* Von Nostrand Reinhold, New York, NY.

National Research Council. 1995. *Wetlands: Characteristics and Boundaries.* National Academy Press, Washington, DC.

On-line resources:

Local precipitation data for all areas of the U. S. is available from the Natural Resource Conservation Service at www.wcc.nrcs.usda.gov/water/wetlands.html.

White water lily,
Nymphaea oderata

SURFACE WATER DEPTHS

Season: ______________ Beginning date: ____________

SURFACE WATER DEPTH	1	2	3	4	5	6	7	8	9	10	11	12	13
3.0 ft													
2.8 ft													
2.6 ft													
2.4 ft													
2.2 ft													
2.0 ft													
1.8 ft													
1.6 ft													
1.4 ft													
1.2 ft													
1.0 ft													
0.8 ft													
0.6 ft													
0.4 ft													
0.2 ft													
0.0 ft													

WEEK

PRECIPITATION

Season: ______________ Beginning date: ____________

AMOUNT OF PRECIPITATION IN RAIN GAUGE

3.0 in													
2.8 in													
2.6 in													
2.4 in													
2.2 in													
2.0 in													
1.8 in													
1.6 in													
1.4 in													
1.2 in													
1.0 in													
0.8 in													
0.6 in													
0.4 in													
0.2 in													
0.0 in													
	1	**2**	**3**	**4**	**5**	**6**	**7**	**8**	**9**	**10**	**11**	**12**	**13**

WEEK

STORM EVENTS

Season: ______________ Beginning date: ____________

CHANGE IN WETLAND WATER DEPTH

	0.0 in	0.2 in	0.4 in	0.6 in	0.8 in	1.0 in	1.2 in	1.4 in	1.6 in	1.8 in	2.0 in	2.2 in	2.4 in	2.6 in	2.8 in	3.0 in	3.2 in	3.4 in	3.6 in
3.0 ft																			
2.8 ft																			
2.6 ft																			
2.4 ft																			
2.2 ft																			
2.0 ft																			
1.8 ft																			
1.6 ft																			
1.4 ft																			
1.2 ft																			
1.0 ft																			
0.8 ft																			
0.6 ft																			
0.4 ft																			
0.2 ft																			
0.0 ft																			

RAINWATER IN RAIN GAUGE

GROUNDWATER LEVELS

Season: ______________ Beginning date: ____________

DISTANCE FROM GROUND SURFACE TO GROUNDWATER

0.0 ft													
0.2 ft													
0.4 ft													
0.6 ft													
0.8 ft													
1.0 ft													
1.2 ft													
1.4 ft													
1.6 ft													
1.8 ft													
2.0 ft													
2.2 ft													
2.4 ft													
2.6 ft													
2.8 ft													
3.0 ft													
	1	2	3	4	5	6	7	8	9	10	11	12	13

WEEK

Data Sandwiches

SUMMARY

Information on topography, soils, hydrology, and plants is gathered within a wetland area. Using the same scale for each data set, the information is layered, thus providing a means of correlating the measured factors and examining their interrelationships.

OBJECTIVES

Data gathered on wetland topography, soils, hydrology, and plants will be correlated spatially. The information will be layered like a sandwich to provide a means of comparing the data and determining the interrelationships among these factors.

MATERIALS

- Survey instrument (a level or transit with a tripod)
- An elevation rod
- Line levels
- 50 to 100 foot (15 to 30 meter) measuring tape
- Balls of string
- Yard or meter sticks
- Survey flags or wire flags
- Short wooden posts to permanently mark corners
- Transparency film
- Colored markers (black, blue, red, green)
- Copies of *Hydrology* Student Page
- Copies of *Soils* Student Page
- Copies of *Plants* Student Page

MAKING CONNECTIONS

Geographic Information Systems (GIS) are computer systems that assemble, store, manipulate, and display geographically-referenced information. Simpler systems have long been used for tracking and correlating information. Bus and commuter train routes are matched with city street grids. Plumbing and electrical wiring plans for buildings are layered over building floor plans. Locations of sewer pipes, water pipes, gas pipes, electric cables, telephone cables, and television cables are all correlated with street and building locations to prevent them from being accidentally severed during construction projects. Whether using computer systems or not, geographic layering of information allows new information to be derived from the relationships that become evident.

BACKGROUND

This activity is useful in understanding both natural and planned wetlands. If a planned wetland is being examined, information collected in two earlier activities (*Discovering Dips* and *What Grows There?*, pages 178 and 127 respectively) provides background data for the geographic information system, as does the school site map. When the planned wetland is completed, however, the area should be resurveyed and a new topographic map created. One-foot contour lines are desirable. Information collected the first time this activity is completed provides the baseline from which future annual comparisons can be made.

If a natural wetland is being examined, a topographic map of the area must be obtained or created. Once the topography of the wetland is known, this activity can be completed and repeated annually.

PROCEDURE

Warm-Up

Before digging a three-foot deep hole to plant a tree in a schoolyard, what information is necessary about potential hidden dangers below ground? (Locations of buried power lines and other utilities must be checked.) How are the locations of utility lines determined? (Call the designated telephone number in your area for buried utilities. They will check a geographic information system to determine what hazards you must avoid.)

Activity

CAUTION: If many small groups complete the entire activity, the wetland will be unnecessarily stressed. It would be appropriate for students to work in small groups or teams on different aspects of the activity. For example, each team could collect the hydrologic, soil, and plant data for one transect, then share information to produce the data sandwiches. The topographic base layer and the grid layer will need to be completed first, however.

Grades K-4

This activity does not fit the abilities of most younger students.

Grades 5-12

A. THE TOPOGRAPHIC BASE LAYER

Obtain or create a topographic map of the wetland showing one foot contours. Use the survey procedures in the activity *Discovering Dips* (page 178) or use the line level/meter stick/string procedure in the activity *Finding Living Benchmarks* (page 149).

If permanent structures such as buildings, sidewalks, or roads are near the wetland, use them as boundaries for one or more sides of the study area. If permanent structures are not available, place permanent markers such as short wooden posts at each of the corners of the wetland study area. Measure the distance along each side of the study area; then record the location of the boundary markers and distances between them on the topographic map. This will form the base layer for the information system. Determine the scale. For example, one inch on paper equals ten feet on the ground, or one centimeter on paper equals one meter on the ground.

B. THE GRID LAYER

Set up a grid by extending the measuring tape between two corner posts along one side of the study area. Place a survey flag or small wooden stake at five-feet, ten-feet, one-meter, or two-meter intervals, depending on the overall size of the study area. Do the same on the opposite side, but starting at the same end of the study area. Extend a ball of string across the study area between corresponding markers on each side. Keep the string line straight. Repeat for each pair of markers on opposite sides of the study area. See **Figure 10.4**.

Designate one corner post the A post, and the next corner post (going clockwise) as the A′ post. (A′ is called A-prime.) These posts should already be connected with a string line along one side of the study area. The next string line will connect B and B′, and the one after that will be C and C′. Continue with this labeling until the last two corner posts are linked.

Along each transect line, place survey flags at regular intervals (five-feet, ten-feet, one-meter, or two-meters) to designate stations along the transect.

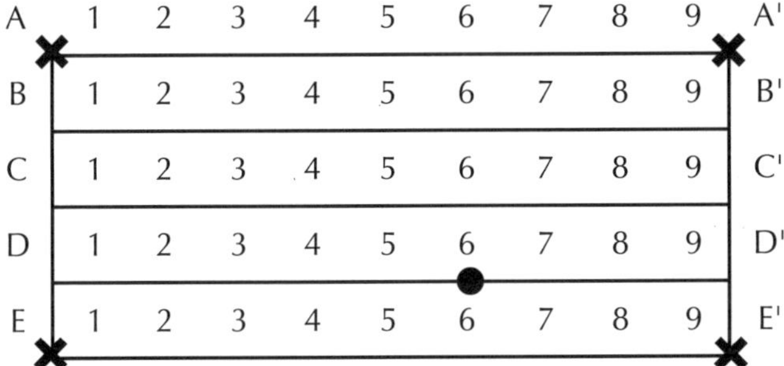

Figure 10.4 Transects and stations are shown with corner posts at A, A′, E, and E′. The marked sample's location is D6.

Back in the classroom, place a sheet of transparency film over the topographic base layer. Trace the marks representing the boundary markers on the topographic layer; this will help ensure that all layers stay properly aligned. Carefully mark the grid on the transparency film using a black marker and a ruler. Pay close attention to the distances measured and use the scale selected.

C. THE HYDROLOGIC LAYER

Observe the hydrology of the wetland site. Complete one copy of the *Hydrology* Student Page for each transect, recording information for every station.

In the classroom, place a second sheet of transparency film over the transect layer. Again, carefully trace the marks representing the boundary markers.

Using the data on the *Hydrology* Student Page and the transect layer as a guide, outline the areas where (a) standing water or tidal water was present using a blue marker. Fill the outlined area with blue color. Outline areas that showed evidence that water had collected there in the past (b-f); then fill the outlined area with blue dots or lines.

D. THE SOIL LAYER
CAUTION: Limit foot traffic through the wetland. Wetland soils are often easily compacted, thereby changing the soil characteristics. Excessive foot traffic can also result in erosion scars that are slow to heal.

Complete the *Soils* Student Page for each station along every transect. If available, use a soil probe to obtain a single sample at each station. Examine colors, textures, moisture content, and smell within six inches (fifteen centimeters) of the surface and between six and twelve inches (fifteen and thirty centimeters) of depth. If needed, refer back to the activity *What Grows There?* (page 127) for more detailed information on techniques.

In the classroom, remove the completed hydrologic layer from the stack. Place a third sheet of transparency film over the transect layer. Carefully trace the marks representing the boundary markers.

Using data from the *Soils* Student Page and the transect layer as a guide, mark the stations having at least two hydric soil indicators within six inches (fifteen centimeters) of the surface. Use a red marker. Connect these locations, then fill the outlined area with red color. Mark the sample locations having at least two hydric soil indicators between six and twelve inches (fifteen and thirty centimeters) below the surface. Connect these locations, then fill the outlined area with red dots or lines.

E. THE PLANT LAYER
CAUTION: Do not remove wetland plants, especially if they are one of a kind. Photograph or sketch plants that are not easily identified. Limit foot traffic in the wetland to avoid crushing wetland plants.

Complete the *Plants* Student Page for each station along every transect. Include all plants touching the string, hanging over the string, or growing under the string or within one foot of the string along both sides of the string. Identify the most common plants. Determine their wetland indicator status. (The National Plants Database provides regional listings of wetland plants for the United States, as well as plant characteristics, profiles, and pictures; see Resources.)

In the classroom, remove the completed soil layer from the stack. Place a fourth sheet of transparency film over the transect layer. Carefully trace the marks representing the boundary markers.

Using the data from the *Plants* Student Page and the transect layer as a guide, mark the sample locations that have at least one obligate wetland species present (OBL) using a green marker. Connect these locations; then fill the outlined area with green color. Mark the sample locations having at least one plant species with the indicator status of facultative wetland plant (FACW). Connect these locations, then fill the outlined area with green dots or lines.

Wrap-Up
Stack the information layers in the following order: the transect layer (hidden from view), the topographic base layer, the hydrologic layer, the soil layer, and the plant layer as in **Figure 10.5**. Do the colored areas

match? If not, what might this indicate? What features do the dotted areas match? What might be the reasons? Are there any indications that the size or shape of the wetland has changed over time? What might cause this to happen?

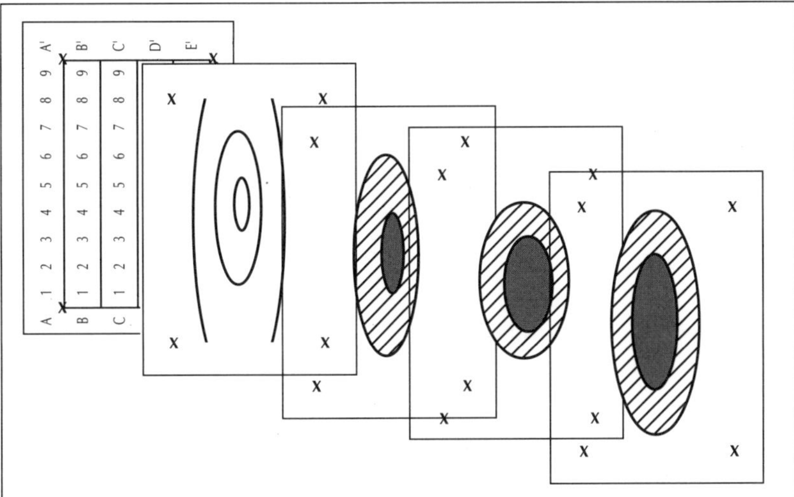

Figure 10.5 Assemble the data layers like a sandwich.

ASSESSMENT

Examine student products. If no relationships are evident, this may be due to poor construction or sloppy work. (The colored areas should not be expected to match up exactly, but some correlations should be evident. The reasons for this will vary from site to site.)

Have students answer the following orally or in writing:

- Describe the interrelationships that seem evident among topography, plants, soils, and hydrology.
- Deduce what changes may have occurred at this site over time.
- Are some plant species found only in specific areas?
- How do hydrology and soils appear to be linked? Why?
- How could these correlations be examined further or tested?

EXTENSIONS

Use the mapping programs in the school technology laboratory to layer the wetland data over the topographic map to produce a true GIS. To produce hard copies, print all layers on transparency film except the base topographic layer.

Use a hand-held Global Positioning System (GPS) to determine positions of the boundary markers and the transects for use in conjunction with the computer mapping programs.

Survey the watershed above and below the wetland for land uses. On a map of the watershed, shade the areas with a different color for each land use. Be sure to include a key. Which land use activities above the wetland will affect the wetland? Will the wetland be able to buffer the areas below it in the watershed? How can the wetland be monitored for changes caused by land use?

Repeat the transects annually during the same time of year. Compare with previous years to determine what changes are occurring. If appropriate, calculate the rate of change in wetland size. Our understanding of wetlands is evolving based on knowledge gained from many years of accumulated data.

RESOURCES

Geographic Information Systems. U.S. Geological Survey. (pamphlet) 1-800-USA-MAPS.

Maryland's Coastal Bays...An Electronic Atlas. Geographic Information Services Division, MD DNR, Annapolis, MD. (demonstration disk)

Some helpful on-line resources:

U.S. Geological Survey gives information on GIS: info.er.usgs.gov/research/gis/title.html

Environmental Systems Research Institute, Inc. has free software and information on GIS: www.esri.com

Geoplace.Com provides links, free software, and the history of GIS: www.geoplace.com

GISPortal.com links with great GIS net sites grouped by state and other criteria: www.hdm.com/gis3.htm

National Imagery and Mapping Agency has a students' corner: www.nima.mil

National Plants Database provides the Wetland Indicator Status of each wetland plant species for each region of the United States, as well as characteristics, pictures, and a profile: plants.usda.gov/plants

Beaver,
Castor canadensis

HYDROLOGY

Transect ______________

When the following characteristics are present, place an X in the column for that station.	Station Numbers									
	1	2	3	4	5	6	7	8	9	10
a. Standing water or tidal water										
b. Depression where water might collect										
c. Low spots with mud or dried mud cracks										
d. Water stains on trees or other vegetation										
e. Vegetation mud-stained from flooding										
f. Gullies or signs of water erosion										

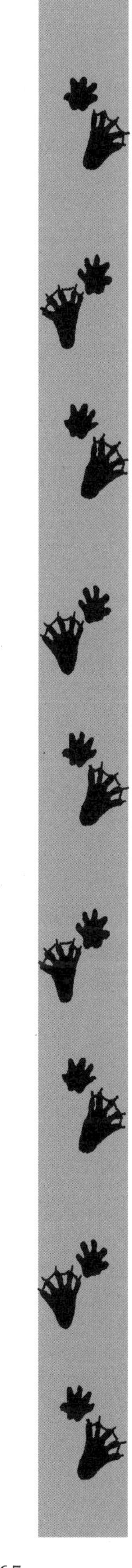

SOILS

Transect ______________

When the following hydric soil indicators are present, place an X in the column for that station.	Station Numbers									
	1	2	3	4	5	6	7	8	9	10
Within 6 inches (15 cm) of the surface:										
Soil is wet.										
Soil is green, dark gray, brown, or black.										
Soil shows mottling or contain oxidized rhizospheres.										
Soil colors are gleyed.										
Soil forms a ribbon at least 2 inches long (5 cm).										
Soil has a rotten egg smell.										
More than 6 inches, but less than 12 inches from the surface (15-30 cm):										
Soil is wet.										
Soil is green, dark gray, brown, or black.										
Soil shows mottling or contain oxidized rhizospheres.										
Soil colors are gleyed.										
Soil forms a ribbon at least 2 inches long (5 cm).										
Soil has a rotten egg smell.										

PLANTS

Transect ______________

When wetland plant species are present, place an X in the column for that station.	Wetland Status	Station Numbers									
		1	2	3	4	5	6	7	8	9	10
Tree species:											
Shrub species:											
Emergent species:											
Aquatic species:											

11 WATCHING WETLAND WILDLIFE

ACTIVITIES

Wetland Journeys

Mammal Marks

Beautiful Birds

Herp Search

Bugs and Butterflies

Aquatic Animals

Wetland Soup

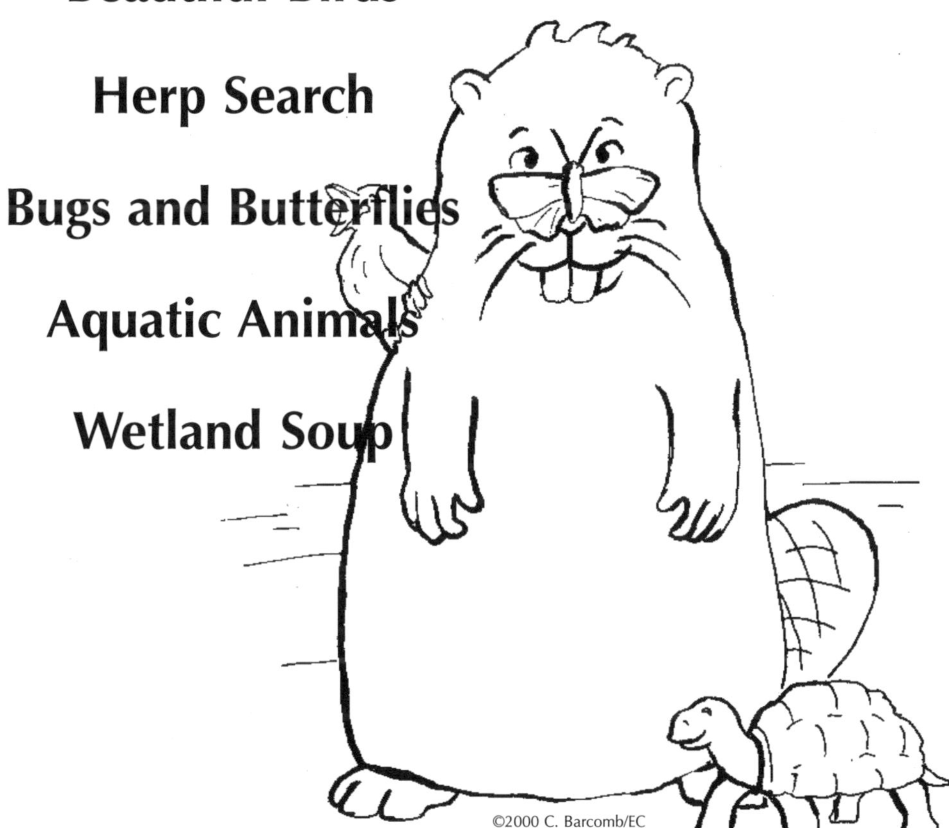

Wetland Journeys

SUMMARY

Observations of natural events, activities, flora and fauna have been recorded in journals for hundreds of years. In our fast-paced world, observations recorded in a handwritten journal allow time for reflection about ourselves and the natural world.

OBJECTIVES

In a handwritten journal, students will record detailed observations of the natural world through words and illustrations.

MATERIALS

- Bound composition books
- Pens or pencils
- Resealable plastic bags

MAKING CONNECTIONS

Observations are made by using our senses. While recording observations in a journal, we begin to organize, analyze, and synthesize them, comparing them with previous knowledge or observations, noting similarities, differences, and details that we may not have paid attention to otherwise. The ability to record observations is a skill that may be applied throughout our personal lives, such as in diaries, baby albums, health and veterinary records; and in many areas of employment, such as in boat captains' log books, in truck drivers' record books, in hairdresser note cards, and in scientists' research notebooks.

BACKGROUND

Humans of the Stone Age painted their observations of bison and other game animals on cave walls. Naturalists such as Aldo Leopold, John Muir, and Henry David Thoreau recorded their observations in journals. Journals help us to understand ourselves and the world around us. Occasionally we have an opportunity to view journals from the past. These allow us to note the changed and unchanging aspects of the natural world (such as wetlands that have been drained and filled), as well as those that have remained the same (the horseshoe crab).

PROCEDURE

Warm-Up

Close your eyes and remember the last bird you saw. Open your eyes, and either sketch the bird or write a description of the bird. Now look at a large picture or poster of a bird, and sketch or write a description of it. Which sketch or description had the most detail? Why was it easier to sketch or describe the second bird? If we sketch or describe creatures when we see them, or soon afterward, we are able to record in more detail and with more accuracy. Journals are memory aids.

Provide students with examples of journal entries by naturalists from the region.

Activity

Grades K-4

Designate one table in (or area of) the classroom as the Nature Table, and explain that this will be a place to display natural (not man-made) objects. Take students on a walk through a natural area or even the schoolyard. Allow them to collect objects to place on the Nature Table. As objects are placed on the table, students may record where and when the objects were found and details about the objects in a notebook kept on the table. Objects for the table might include a rock, a feather, or a leaf. Based on what is recorded in the notebook, allow students each week to show and describe their objects to the class. Objects on the Nature Table should change over time.

Grades 5-12

Go to an outdoor area large enough for everyone to sit, and observe nature without talking to each other. Each person should have a composition book that will be used exclusively for journal entries. Use either a pen or pencil. Pencils lend themselves to sketching, but use whatever is comfortable. A resealable plastic bag is useful in protecting and storing journals and pens or pencils.

The goal is to observe natural objects and events, and record them. The style may include sketches, technical details, and personal reflections. There should be no set length for the assignment. Curiosity and observational skills need encouragement, and suffer if a strict format is required. Sketching should be encouraged, since more detail is usually observed. Sometimes only a portion of a organism will be sketched, such as a beak, tail, or flower. Even motion may be sketched.

To begin, record the date, time of day, location, and weather; this allows for comparisons over the seasons and years. Record observations of whatever you feel is interesting, and note how you are affected by those observations. Look and listen for signs of creatures, large and small. A plant may be observed, but there are also many creatures that inhabit plants. The longer everyone remains still and quiet, the more likely creatures will be to return to their normal daily activities. Insects will search for favorite plants or prey, birds will search for insects or berries, chipmunks will forage, turtles will sun themselves, and rabbits will venture out.

Wrap-Up

Bring everyone back together as a group, and have students name the organisms they observed and what they felt were the most interesting details about them.

ASSESSMENT

- Are detailed observations recorded in the journals?
- Are the details sufficient that the organism (or an important part or action) is recognizable by others?
- Do the journal entries employ several senses?

EXTENSIONS

Continue making journal entries. Whether observing deer in a natural area, butterflies in a park, insects in a yard, or birds from a window---students can practice their skills of observation frequently, even daily, and provide valuable information about the patterns and rhythms of nature.

RESOURCES

Cornell, J. 1989. *Sharing the Joy of Nature.* Dawn Publications, Nevada City, CA.

Eastman, J. 1995. *The Book of Swamp and Bog.* Stackpole Books, Mechanicsburg, PA.

Hinchman, H. 1991. *A Life In Hand–Creating the Illuminated Journal.* Peregrine Smith Books, Layton, UT.

Hudgens, D. "Outdoor Diaries: Writing Nature's Biography". *The Washington Post Weekend.* March 13, 1998. Washington, DC.

Johnson, C. 1990. *The Sierra Club Guide to Sketching in Nature.* Sierra Club Books, San Francisco, CA.

Leslie, C. W. 1995. *Nature Drawing–A Tool for Learning.* Kendall/Hunt Publishing Company, Dubuque, IA.

Leslie, C. W. 1995. *The Art of Field Sketching.* Kendall/Hunt Publishing Company, Dubuque, IA.

Canada geese,
Branta canadensis

Mammal Marks

SUMMARY

Mammals are elusive creatures, which is why we thrill to see a deer appear at the edge of a wet meadow or a fox dash across a road at night. Because we observe wild mammals less often than other creatures, we delight when we do catch a glimpse of them. Several methods are presented that will improve the odds of finding wetland mammals or, at least, of finding evidence of their presence among us.

OBJECTIVES

Students will locate and identify tracks or other signs of wetland mammals. Students will speculate about what was happening when the sign was made.

MATERIALS

- Transparency of *Tracks* Student Page
- Field guides to mammals and tracks
- Plaster of Paris
- Resealable plastic bags
- A variety of baits
- Brush cuttings
- Live traps
- Copies of *Records of Wetland Mammals* Student Page

MAKING CONNECTIONS

Tracking is one of the oldest skills, one that our ancestors depended upon to find food. Now that we hunt our meat in the isles of grocery stores, where we find it neatly wrapped in plastic packages, we seldom practice our tracking skills. Yet, with just a little effort we can begin to see the trails, signs, and tracks of the wetland mammals still among us.

BACKGROUND

Mammals generally have fewer offspring and spend more time raising their young than other groups of creatures. Mammal populations, therefore, naturally tend to be small when there is equilibrium within an ecosystem. The extensive hunting and trapping of our ancestors upset that equilibrium. Deer were completely eliminated along most of the East Coast, because they were an easy source of food for early settlers. Most predators (such as coyotes, wolves, bobcats, and bears) were also eliminated, because settlers feared them and the damage they could do to domestic livestock. Muskrats and beavers were trapped for their fur, which eventually became clothing.

Most mammals that live among us are nocturnal and, therefore, seldom seen. With hunting and trapping pressures reduced, some are extending their ranges. Coyotes and bobcats have become residents of suburban and rural areas, protected in part by their nocturnal habits and by their resemblance to pet dogs and to house cats. Bears have been captured in New York City's Central Park, and mountain lions have been recorded in the Virginia suburbs of Washington, DC. Muskrat, beaver, river otter, and wild mink (or their signs) can be seen in nearly every wetland large enough to support them. Because of this, maps in field guides are highly unreliable. Check with a naturalist in your area to determine which creatures are likely to be present.

PROCEDURE

Warm-Up

Look at the *Tracks* Student Page and tell a story about what happened. How many animals made the tracks? (Two) Where they the same kind of animal? (No) What are the possible kinds or species of animals? (Raccoon and dog) What may have happened? (The raccoon walking in lower left is spotted by a dog walking in lower right; the dog gives chase and both run at top.)

Activity

All grades

Try some or all of the following techniques to locate wetland mammals. Keep personal and class lists of the mammals sighted, the date, time of day, location, and weather on *Records of Wetland Mammals* Student Page.

A. CASUAL OBSERVATION (Individual or small group)
In the early morning and in the evening just before dark, mammals such as deer come out to feed on lush marsh grasses. At dusk, first the hawks, then the owls will prey on mice, shrews, and rabbits that live in the marsh. Beaver often are active on foggy days. Often just before and just after a storm, creatures are especially active. Learn the time of day when mammals in your area are most likely to be active. Then quietly watch from a hillside overlooking a wetland, use a blind for cover, or walk quietly along a wetland trail. Sightings are more likely to occur if everyone is quiet so that the animals are not spooked, and if the wind (no matter how slight) is blowing from the wetland to the people so that animals are not alerted to human presence before they can be seen. Remember that animals' senses are much keener than ours. Note the location and activity of each creature.

B. BECOMING PART OF THE HABITAT (Individual or small group)
Find a location in the wetlands that has a broad view of the area. Sit down or lean against a tree, keep still, and wait. As people become part of the habitat, creatures return to their normal activities and become more visible. Identify and record any creatures encountered, what they were doing, and where they were seen.

C. SIGN SEARCH (Individual or group)
While slowly following a wetland trail, watch for side trails or less frequently used runs made by mammals as they pass through vegetation. Occasionally pause and crouch down to see if smaller tunnel-like trails are visible in the underbrush. Search for tree rubs where male deer have scraped tree bark as they rub the velvet from their new antlers. Look for

beds of matted grasses where creatures have rested, and other well-worn depressions in the soil that indicate the size and shape of a creature that has slept there. Felled trees and branches with tooth marks indicate the possible presence of beavers. Look for small piles of leaves and muck near the water's edge that beavers and muskrats construct to mark territory. Watch for scat (dung, feces, or droppings) and carcasses from interrupted meals; both tell what an animal has been eating and, therefore, where it has been. Look for muskrat lodges, beaver lodges, beaver dams, and den holes. Along banks, look for mud or snow slides made by river otters at play. On a map of the wetland, mark where mammal signs were found, who left the mark, and what the creature was likely to have been doing when the sign was made.

D. BEAVER SEARCH (Group activity)
Beavers are the original wetland engineers. When colonizing an area, beavers first dig a burrow in the bank of a stream or creek, then fell trees and shrubs that are then laid across the stream channel, creating a dam. The dam height is increased through the addition of mud, rocks, and sticks, causing water to back up behind the dam. As the flooded area expands, beavers gain access to more trees. As the water rises, mud and branches are piled on top of the original bank burrow, creating a lodge that sometimes becomes surrounded by water.

Beaver signs include:

- barkless sticks with teeth marks showing where beavers have fed on the inner bark or cambium layer;
- pointed stumps with teeth marks, indicating the tree was felled by a beaver and then moved (often floated) to a dam or lodge;
- a stick covered lodge along a bank or surrounded by water;
- a stick built dam with teeth marks on the ends of the sticks;
- dead or dying trees in an area that is becoming a marshy meadow; and
- beaver tracks with the characteristic tail drag mark between them.

Beaver dams are often visible in aerial photographs and are sometimes marked on maps. In ten to twenty years, a beaver colony will exhaust the food supply of an area and move on, but the changes the beavers have made in the habitat will persist for many years.

Do beavers live in your area? Are there signs of beavers in or near your wetland? If so, what signs? How long have beavers been in the area? How many are current residents? How have the beavers changed the local habitat? Check with your state Department of Natural Resources for help in locating beaver lodges in your area.

E. BAITING A SAND BOX (Group activity)
In an accessible area of the wetland, use boards to form a square on the ground, with one edge of each board partially buried in the ground. Use wetland muck or other soil that will hold a shape when stepped on, or a bag of fine sand to fill the square; then smooth it. Place a rock, small board, or piece of log in the center as a bait stand. Place on the stand nuts, a piece of apple, an ear of corn, a piece of orange, honeysuckle flowers, or whatever bait an animal might eat. Then leave. Check

the following day to see if the bait is missing. If it is, look in the sand for tracks made by the creatures that came during the night. Using an animal tracks guide, try to determine the type of animal. Keep a log of the baits used and of the creatures that fed on the bait.

Under these controlled conditions, mammals often leave good tracks that can become a permanent record of the creature. Mix plaster of Paris (carried in a resealable plastic bag) with enough water (a handful or so from the wetland or a water bottle will do) to form a paste. Reseal the bag and knead it; pour the mixture into and slightly beyond the track. After the plaster has hardened, lift up the mold of the track. Be sure to smooth the ground so that as new tracks are formed they can be easily seen.

When the plaster is fully dry, gently brush the sand off the mold using an old toothbrush or small paintbrush. Some details of the track, such as claw marks, may be more obvious on the mold than in the sand. After molds of several tracks are collected, you can make your own animal tracks in a sandbox and create a story about what might have happened.

F. TRACKING (Individual or small group)
The soft mud that is generally a part of a wetland is an excellent surface on which to find animal tracks. Walk around the wetland in search of tracks large and small. When tracks are found, observe their shape, the presence or absence of claw marks, evidence of tail dragging, the depth of the tracks, and whether the tracks are fresh and moist or old and dry. Look for other similar tracks nearby. Observe the trail pattern; it can be just as important in identifying a creature as the prints are. Measure the distance between prints (the stride) and the trail width. (See **Figure 11.1**.) The size of the print itself may be distorted by soft mud, but the print pattern (the relationship among the tracks made by all four feet) often allows accurate identification of an animal. Record the name of the wetland creature that created the trail or track, and look for clues that tell what the creature was doing at the time the tracks were made.

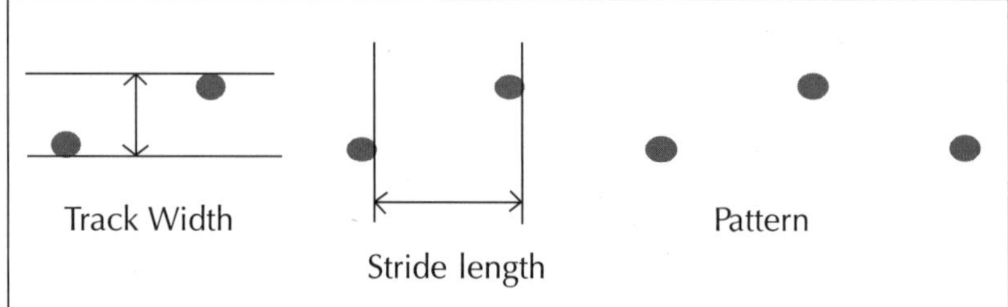

Figure 11.1 Characteristics that help identify tracks are track width, stride length, and pattern.

G. BRUSH PILES (Group activity)
Underbrush provides a safe haven for small mammals where they can raise families and hide from predators. If the wetland has very little underbrush, small brush piles may be constructed of grass cuttings, small branches, and discarded Christmas trees. In time, alarmed creatures may be seen running in the direction of the brush pile for safety. Frightened creatures will often pause in secondary hiding places near their

brush pile refuge rather than lead a predator to the entrance, because that would lessen its value as a safe haven. Look for tracks near the brush pile, trails leading to the brush pile, and den holes in the brush pile. Record what is observed.

H. LIVE TRAPS (Individual or small group)
Live traps come in a variety of sizes and are helpful in determining what creatures inhabit an area. Traps can be baited and placed where creatures are known to travel. If traps are placed out of the sun (so that trapped creatures do not dehydrate), away from water (so that trapped creatures are not likely to drown in a sudden rainstorm), and checked regularly (so that trapped creatures do not starve or die of exposure), then live traps can be a safe means of determining what mammals live in an area. When a creature is trapped, record the animal's species name, its approximate size, and the date and time. Carefully release the animal where it was found.

Wrap-Up

According to the *Record of Wetland Mammals*, how many different species of mammals live in or use the wetland? Which species is the most abundant? How do various mammals use the wetland? Are some areas of the wetland used more than others? Where? Are some mammals active only at certain times of day? If so, which ones are active at which times? Keep individual records for the entire year, but enter class records in the "Wetland Log Book" for year-to-year comparisons.

ASSESSMENT

Have students do the following activities:

- Draw, as accurately as possible, the tracks of a particular species of animal.
- Describe what signs other than tracks might reveal the presence of this creature?
- Draw two sets of tracks that intersect in some manner.
- Write a short story about what was happening when the tracks were made.
- In the story, describe what and how each creature sensed. What did each creature feel, and why?

EXTENSIONS

Continue looking for marks that wetland mammals make as they go about living their lives. Keep a continuing record of the mammals present in and near the wetland, and their habits. Try to recognize the marks made by individual mammals, by noting details such as a scar on a foot pad, a limp, or an unusual size. It may help to name the individual animals.

Go on-line with the University of Washington (www.fish.washington.edu/naturemapping) or the state of Virginia (www.dgif.state.va.us/wildlifemapping) to map species' ranges and to work with real data to investigate wetland creatures.

RESOURCES

Brooks, R. P., D. A. Devlin, and J. Hassinger. 1993. *Wetlands and Wildlife.* Pennsylvania State University, University Park, PA.

Brown, T. Jr. and B. Morgan. 1983. *Tom Brown's Field Guide to Nature Observation and Tracking.* Burkley Books, New York, NY.

Burt, W. H. and R. P. Grossenheider. 1976. *The Peterson Field Guide Series: A Field Guide to the Mammals.* Houghton Mifflin, Boston, MA.

Fadala, S. 1989. *Basic Projects in Wildlife Watching.* Stackpole Books, Harrisburg, PA.

Miller, D. S. 1981. *Track Finder.* Nature Study Guild, Rochester, NY.

Murie, O. J. 1974. *The Peterson Field Guide Series: A Field Guide to Animal Tracks.* Houghton Mifflin, Boston, MA.

Rezendes, P. 1995. *Tracking and the Art of Seeing: How to Read Animal Tracks and Sign.* Camden House Publishing, Inc., Charlotte, VT.

Stall, C. 1989. *Animal Tracks of the Mid-Atlantic States.* The Mountaineers, Seattle, WA.

Vaughan, T. A. 1972. *Mammalogy.* W. B. Saunders Company, Philadelphia, PA.

Zim, H. S. and D. F. Hoffmeister. 1955. *Mammals.* Golden Press, New York, NY.

On-line resources:
U. S. Fish and Wildlife Service at www.fws.gov.

Meadow jumping mouse,
Zapus hudsonius

TRACKS

RECORDS OF WETLAND MAMMALS

Date: ________________ Collection/Observation Method: ____________________________

Weather: ____

Time	Location	Species	Activity

Beautiful Birds

SUMMARY

Both resident and migrating birds are attracted to water. Several ways are presented to observe these beautiful wetland creatures.

OBJECTIVES

Students will plan ways to attract wetland birds to identify them and observe their behavior.

MATERIALS

- Tapes of bird songs
- Field guides
- Binoculars
- Bird feeders and seeds
- Materials for nest boxes or platforms
- Birdcalls and decoys
- Materials for a blind
- Lures and bait, such as fruit or nuts
- Copies of *Records of Wetland Birds* Student Page

MAKING CONNECTIONS

Bird watching is often a lifelong experience, whether approached formally or informally. For many people, it is both recreation and exercise. Bird watching is an activity that is available to both the physically fit and the handicapped, the young and the old. Birds are found on every continent and in every ecosystem. They can be seen in remote wilderness areas and in the most crowded cities. Some are kept in cages, but most are wild and free. To watch birds, all we need do is look through a window or step outside. The birds are there; all that is required of us is our attention.

BACKGROUND

Birds may migrate thousands of miles each year across continents and ecosystems as they move between summer and winter habitats. During these journeys and at both destinations, they will need food, water, and a safe place to rest. Water is an important part of bird habitat, so wetlands are particularly good places to observe birds. The variety and abundance of foods (such as seeds, berries, insects, crustaceans, and fish), the diversity of habitats, and the presence of water attract many different species to wetlands.

PROCEDURE

Warm-Up

Grades K-4

Go outside to the wetland. Sit down, close your eyes, and listen for birds singing. Raise one finger each time you hear a bird sing. When most students have five fingers up, you will be told to open your eyes. Take turns imitating the bird songs that were heard while pointing in the direction from which the song came. Can the birds be seen?

Grades 5-12
Listen to a tape of bird songs that identifies the birds as they sing. Which bird calls have been heard locally? Write down their names; then try imitating their song. If a complete list of bird songs on the tape is prepared ahead of time, a check can be placed next to the names of local birds.

Activity

All grades
Some of the following activities are expensive, and some are labor intensive. Choose activities that suit the abilities and needs of your particular group. Keep personal and class lists of the birds sighted, the time of day, location, date, and weather on *Records of Wetland Birds* Student Page.

A. NATURE WALK (Individual or small group)
In small quiet groups, follow trails throughout the wetland. Using field guides and bird lists for the area, see how many birds can be observed and identified. If available, use binoculars to see more clearly the characteristics necessary for accurate identification of birds.

When a bird has been spotted, observe its behavior. In which of the following activities is the bird engaged?

Hunting for food on the ground
Hunting for food in trees or shrubs
Hunting for food while flying
Approaching or leaving a nest
Sitting on a nest
Singing
Giving an alarm call
Resting or roosting
Flocking
Being aggressive
Flying

Be sure to record your bird identifications and observations.

B. LISTENING POSTS (Good group activity)
In a suitable spot, sit down as a group and listen quietly. When a bird song is heard, students should raise their hands and point to its source without talking. Was the call high in a tree or low near the ground? Was it a mating call, a flock or family chatter, or an alarm call? Male birds often "sing" a territory, that is they sing a song from the highest perch within the territory they claim. This tells other males of the same species to stay away, and females of the same species that they are looking for a mate. Flock and family chatter are usually soft sounds made while feeding or flying together. Alarm calls are sudden, loud, and often accompanied by a flurry of activity.

If the bird can be seen, what kind is it? Use a field guide or local species list. If not readily identifiable, note its color pattern; the size and shape of the legs, wings, tail, and beak; and the song. Record how many birds of each kind are discovered, where they are located, what they are doing, and the time of day.

C. BIRD FEEDERS (Individual or small group activity)
Feeders may be purchased or built. They may be made of wood, PVC pipe, plastic, ceramic, glass, or recycled milk cartons or jugs. The style may be a tube (with or without perches), a tray, a platform, or a house feeder.

Place the feeder in a location that has easy access, so that it can be refilled even in bad weather. Also, be sure that discarded seed shells and bird droppings can be dealt with easily. If squirrels are present, be sure the feeder is at least ten feet away from any tree or branch, since squirrels are excellent jumpers.

In some areas, it may be necessary to protect the feeder from uninvited guests. A cone-shaped baffle (pointed end up) placed below the feeder if it is on a post, or above the feeder if it is suspended, will prevent most four-footed creatures from gaining access. A length of wide diameter downspout material surrounding a post feeder is also effective. Tube style feeders can be protected by metal mesh.

If several feeders are used, try placing a different feed in each one to bring in a variety of birds. Sunflower seeds are a favored food and are particularly attractive to heavy-billed birds like cardinals and blue jays. Smaller seeds like thistle will attract goldfinches. Corn is favored by ducks, geese, and quail. Wild bird seed mixes often involve a lot of waste as birds kick out everything else to get to the seeds they prefer.

Make sure all feeders are filled regularly; birds easily become dependent on them as a food source. Also, be sure to keep feeders clean. After severe storms, check to make sure feeders have dry seed in them, as wet seed will quickly spoil. Sunflower and thistle feeders should be cleaned monthly with hot soapy water and some bleach to prevent the spread of avian diseases.

If feeders are visible from a window, keep a log book there so observations can be readily recorded. List date, time of day, weather, species seen, and how many.

D. NEST BOXES AND PLATFORMS (Small group activity)
Nest boxes for wood ducks and platforms for ospreys are often built and placed in wetlands because of steep declines in their population numbers. Fortunately, both species are making a comeback. Monitor existing boxes and platforms to determine if they are used by their intended occupants.

If a suitable location is available with written permission from the property owner, nest boxes and platforms can be built and installed by students. An osprey platform should be a wooden square 48 by 48 inches with a rail around the edge to help keep the nest in place during storms. It is usually mounted high above a marsh on a post the size of a telephone pole. Strength, ingenuity, and careful planning will be required to set a platform in place.

The floor of a box for wood ducks is ten by eighteen inches, and the box height is up to twenty-four inches. The entrance hole is four inches wide, and twelve to sixteen inches above the floor of the box. The box is mounted on a post ten to twenty feet above the water in a wooded wetland area. Usually a cone baffle made of aluminum is placed on the mounting post below the box to prevent raccoons and snakes from preying on the eggs.

Record dates of first occupancy in the spring, the species of the occupants and the number of offspring produced.

E. BIRDCALLS, DECOYS, AND BLINDS (Small groups)
Many bird species are gregarious and seek the company of their own kind. Ducks and geese will respond to calls (commercial birdcalls are available in sporting goods stores) by landing where they see and/or hear other birds. If appropriate birdcalls and decoys are available, students might attempt to call birds into an open area where corn has been scattered. This will only work, however, if students can remain quiet and hidden from view. To do this, a blind may be necessary. With the help of knowledgeable parents or community members, a blind can be constructed of cornstalks, brush, or boards near the wetland. This makes an excellent spot for observing other wetland creatures as well. Be sure to record what happens.

Birdcalls are available for songbirds and can usually be purchased in the gift shops of nature centers. One such device is made of wood with a small dowel in a center hole. When the dowel is twisted, a variety of bird-like sounds are produced that may attract curious birds. With practice, the call may be used to "talk" to specific birds. Remember to record your observations.

F. LURES AND BAITS (Small or large groups, depending on location)
Toy birds, toy rabbits, and other toy creatures that are appropriately colored may be used as lures to attract hawks, owls, and other wetland predators. To photograph predators in action, place the lure in an open area that is easily viewed from a building or a blind. Do not attach the toy to the fence, stump, or ground, however, as this may cause injury to a fast-moving predator when it swoops in to carry off the lure. Do be prepared to lose the toy or see it destroyed when the predator learns of the deception.

Other baits such as apple and orange slices, or peanuts work well with some birds and other wetland creatures as well. Keep records of which creatures like which baits; be aware that this may change with the seasons.

Wrap-Up

Carefully compile the data on birds from all sources. Post the data on the blackboard, a poster, or an overhead transparency so that everyone can see it. How many species of birds were seen or heard? Which species was the most common? Which was the least common? Which areas of the wetland had the greatest variety of birds? Which birds are most active in the morning? Which ones are more active in the afternoon?

Keep individual records for the entire year, but enter class records in the "Wetland Log Book" for year-to-year comparisons. How have the species diversity, population numbers, and use of the wetland by the birds changed through the seasons? What changes have occurred over the years? Why?

ASSESSMENT

Based on data collected, have students do the following:

- Draw a time line showing the activities of the three most common birds in the wetland throughout the day.
- Construct a graph or time line showing the time of year during which each species of bird is present in the wetland.

EXTENSIONS

Continue monitoring the wetland birds over several years. Locate another wetland nearby, and determine if the same birds are present there. If there are different species of birds present, determine why.

Participate in Classroom Bird Watch, Project Feeder Watch, Cornell Nest Box Network, or Project Pigeon Watch by reporting bird counts on-line at www.birdsource.cornell.edu.

Begin a life list of birds sighted and identified.

RESOURCES

Borror, D. J. 1967. *Common Bird Songs.* Dover Publications, Inc. NY. (Tape and booklet).

Brooks, R. P., D. A. Devlin, and J. Hassinger. 1993. *Wetlands and Wildlife.* Pennsylvania State University, University Park, PA.

Fadala, S. 1989. *Basic Projects in Wildlife Watching.* Stackpole Books, Harrisburg, PA.

Lingelback, J. 1986. *Hands-On Nature.* Vermont Institute of Natural Science, Woodstock, VT.

Peterson, R. T. 1947. *The Peterson Field Guide Series: A Field Guide to the Birds.* Houghton Mifflin, Boston, MA.

U.S. Fish and Wildlife Service. 1997. *For the Birds.* U.S. Fish and Wildlife Service, Washington, DC.

On-line resources:

American Birding Association; contests, camps, scholarships, links; www.americanbirding.org.

Baltimore Bird Club; bird feeders, bird houses, links; www.bcpl.lib.md.us/~tross/by/backyard.html.

Bird Monitoring; marsh bird monitoring, raptor monitoring, maps of bird abundance during migration; www.mp1-pwrc.usgs.gov/birds.html.

Cornell University; bird songs and images, data collection, monitoring programs (Project Feeder Watch, Cornell Nest Box Network, Project Pigeon Watch) and curricula (Classroom Bird Watch); www.birdsource.cornell.edu.

Ducks Unlimited; wetlands and waterfowl; www.ducks.org.

Journey North Program; track bird migrations; www.learner.org/jnorth/jnorth.html.

National Audubon Society; species lists and profiles, classroom kits and projects; www.audubon.org.

Ornithological Council; state, national, and international organizations; bird lists, species information, identification tips; www.nmnh.si.edu/BIRDNET.

Partners In Flight; teaching materials; www.PartnersInFlight.org.

U.S. Fish and Wildlife Service; information on migratory birds and waterfowl; www.fws.gov.

Belted kingfisher,
Ceryle alcyon

RECORDS OF WETLAND BIRDS

Date: ________________ Collection/Observation Method: ____________________________

Weather: ____

Time	Location	Species	Activity

Herp Search

SUMMARY

Herpetology is the study of reptiles and amphibians, two groups of creatures that, until a century ago, were classified as a single group known as herptiles. Ways to search for and identify these elusive wetland creatures will be presented, as well as one method used by researchers to trap them.

OBJECTIVES

Students will locate herptiles by sight or sound, analyze their characteristics, and classify them. Patterns of habitat use, seasonal and daily activities, species diversity, and population sizes may be determined through analysis of the data collected.

MATERIALS

- Copies of *Frog Song Cards*
- Vented plastic containers
- Resealable plastic bags
- Small dip nets or minnow nets
- Copies of *Records of Wetland Herptiles*

MAKING CONNECTIONS

Amphibians are tied to wetlands because of their mode of reproduction. And since chemicals can easily penetrate their moist skin, they are particularly sensitive to chemical pollution. When students near the Great Lakes discovered numerous frogs with deformed legs, researchers were prompted to study possible links between frog deformities and chemical pollution of the environment. Researchers have since determined that at least some of the deformities were caused by parasites that disrupt normal development of eggs into frogs. But just as canaries were once used to warn of the buildup of lethal gases in deep coal mines, amphibians may be warning us of a chemical buildup in our environment.

BACKGROUND

Amphibians are characterized by moist glandular skin that dries out easily; therefore, they are generally intolerant of salt water. They live the first part of their lives in water and breathe with gills; they spend the second part of their lives on land and breathe with lungs. Because amphibians are unable to generate their own heat, they must hibernate during cold weather in the mud at the bottoms of wetlands or on land below the frost line.

During the spring and summer period of courtship, male frogs call female frogs to come to the water for mating and egg laying. Each species has its own song, so this is one means of identifying frogs, even if they cannot be seen. After a mating ritual, the female lays her eggs in water, where they are fertilized by the male frog. The eggs develop in the water, then hatch into tadpoles that eat algae. The tadpoles then change (metamorphose) into adult frogs that eat insects.

Frogs, toads, and salamanders are all amphibians. They all lack claws, but there are some differences.

- Salamanders have tails and are voiceless; they are found under debris at the water's edge.
- Treefrogs have pads on their toes and webbing between their toes, so they are good climbers.
- True frogs have webbing, but no pads on their toes, so they are good swimmers.
- Toads have no pads and no webbing, so they stay on the ground except when breeding in the water.

Reptiles found in wetlands include lizards, skinks, turtles, and snakes. The skin of reptiles is covered with protective scales and their toes have claws. Like amphibians, they have no internal way of regulating their body temperature, but they often warm themselves by sitting in the sunshine on rocks, stumps, or logs in or near water.

Reptiles lay leathery eggs in holes that the female has dug on land; she then covers the eggs and leaves them. Often these nests are raided by predators, such as raccoons. If this happens, the remnants of the egg shells will be scattered around the nest site. If the reptiles hatch and leave the nest, the egg shells will stay buried in the ground and will not be visible. When reptiles hatch, they look like their parents, except that they are smaller and sometimes have protective coloration.

PROCEDURE

Warm-Up

Copy and cut out enough *Frog Song Cards* so that each person will receive a card. Mix the cards and distribute them to students. All cards are secret (do not show them to anyone) and no sounds are allowed except frog songs.

At a given signal, male frogs should begin calling; females should listen to locate an appropriate partner. Depending on how clearly and loudly they call, some male frogs will have many partners, some will have none. Continue until all female frogs have found partners. Based on frog species, everyone has now been assigned to one of four groups for small group activities.

Activity

All grades

Select some or all of the following techniques to monitor herptile populations. Remember to log all herptile sightings and counts on *Records of Wetland Herptiles* Student Page. Be sure to keep these records in a "Wetlands Log Book" for use in determining patterns of activity, habitat use, species diversity, and population sizes. These records will document long-term changes.

A. CASUAL OBSERVATIONS (Individual or small group)
The day after a rain, take a quiet walk around the wetland. Pause occasionally to look and listen. Record sightings of herptiles in individual nature journals or on data sheets. Be sure to indicate the date, time, weather, location, and species descriptions. Also record the activity in which the creature was engaged (egg laying, basking, swimming, mating, calling, running away). Summarize the information for class records.

B. FROG SONGS (Group activity)
Listen to a recording of frog songs (see Frogwatch USA under Resources) and/or have a naturalist teach some of the songs of local frogs. Once a month from February to September, go to the wetland and very quietly listen for frogs. Identify and record the names of as many species as possible. Frog songs may also be recorded in the wetland, then listened to in the classroom for identification.

C. EGG MASSES (Group activity)
Once a month from February to September, check for egg masses along the water's edge. If an egg mass is found, gently scoop up some of the eggs with a fine mesh net or plastic container. Using a field guide and local species list, attempt to identify the species that produced the eggs. Try not to touch the eggs with your hands, since this may cause disease or injury. Gently return the eggs to the water where they were found in as short a time as possible. Mark the location of the egg cluster on a map of the wetland (along with the species name).

D. BASKING TURTLES (Individual or small group activity)
Turtles often bask in the sun at regular times each day, usually between 10:00 AM and 2:00 PM. If an elevated area is available that allows a view of basking sites (such as a roof or a blind), then this technique can be used to determine the minimum population size and species diversity of turtles.

First determine which turtle species occur locally by checking with a nearby nature center. Then locate an observation post that allows a view of several basking sites, but is far enough away that turtles will not be frightened into the water. Binoculars or a spotting scope are helpful in identifying turtles at a distance. Make several counts during a thirty-minute interval. Record the date, time of day, weather, and number of each species seen basking. Repeat the observations on a weekly or monthly basis to improve accuracy of the counts.

E. TURTLE NEST COUNTS (Group activity)
During summer, female turtles briefly visit dry land to lay their eggs. Turtles select a suitable location for the nest, dig a hole, lay the eggs, then cover and camouflage the nest. Intact turtle nests are not likely to be located by humans unless a female is discovered while she is laying eggs. Turtle nests, however, are frequently preyed upon by raccoons, skunks, and opossums who dig up and eat the eggs. Eggshell fragments scattered about the nest site remain visible for several weeks.

On a weekly or monthly basis, walk about the wetland looking for signs of turtle nests. Destroyed nests can be counted, their locations mapped, and possibly the species identified by looking carefully at the shell fragments. Collect the eggshell fragments to prevent counting the same eggshells again on a later survey.

F. ANNUAL CENSUS (Group activity)
In small groups, search assigned areas in and around a wetland during a single day in May or June when the herptiles are most active (the day after rain is best). Gently turn over logs (return them after looking!), rake through leaf litter, check stream beds and banks, and carefully check the water's edge. Attempt to capture, identify, and measure all herptiles, then record their location on a map of the wetland area. Use small nets to capture the creatures. If handling the animals, make sure hands are first moistened with non-chlorinated water. If creatures are laying eggs, leave them alone! Attempt to identify them at a distance and record the nest location on the map.

Once captured, the creatures may be placed in a small plastic carrying container so that they will not be accidentally crushed while they are being identified. Keep the creatures in the shade so they do not become overheated or dehydrated.

Length may be measured using a ruler. Measure from the tip of the snout (nose) to the tip of the tail for salamanders, skinks, lizards, and snakes. Measure from snout to vent (anus) in frogs and toads. Measure the length of the bottom of a turtle's carapace (shell). To measure wriggling creatures, move them to a small plastic bag so they hold still, but be careful not to smother them! Return all creatures as soon as possible to the same place they were found.

Grades 9-12

The following technique can be very hazardous to herptiles and should not be attempted unless the traps will be faithfully monitored daily!

G. DRIFT FENCES AND PITFALL TRAPS (Individual or small group activity)
If a small group of reliable individuals plan to do a research project on herptiles, excellent directions for construction and use of drift fences and pitfall traps is included in *Save Our Streams: Handbook for Wetlands Conservation and Sustainability*, 1998. The Y-trap suggested is used in general surveys of herptiles. Wetland researchers tend to use a single drift fence with four pitfall traps, two on each side of the fence and two at each end of the fence, to derive information about breeding populations. (See **Figure 11.2**.) If the fence is placed a short distance above a wetland area, herptiles can be intercepted going to and from the wetland, and their direction of travel can also be determined by the traps in which they are caught.

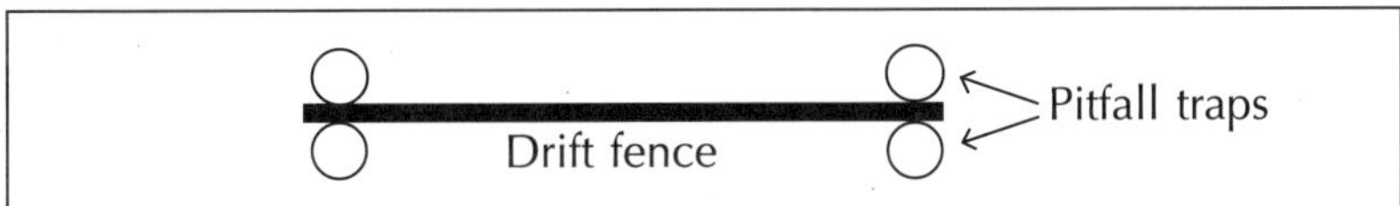

Figure 11.2 A drift fence may have four or more pitfall traps to collect herptiles traveling to and from the wetland.

Wrap-Up

Look carefully at the data collected. Post it on the blackboard, a poster, or an overhead transparency. How many species of herptiles did you find? Which species was most abundant? Which was the rarest? How do these species use the wetland? At what time of day do they use it and for which activities?

If this is a continuing activity, how have species diversity, population numbers, and use of the wetland by these creatures changed over time? Why?

ASSESSMENT

Based on the data collected, have students do the following:

- Draw the life cycles of the most common reptile and the most common amphibian, showing when and how they use the wetland.
- Describe how these herptiles will be affected if the wetland is suddenly drained.

EXTENSIONS

Continue monitoring the wetland creatures over time. Locate another wetland nearby and determine its similarities to and differences from the first wetland. Monitor the herptiles at the second wetland. Determine how the animal populations differ between the two wetlands.

Join the North American Amphibian Monitoring Program. Data that has been collected can be entered on-line at www.mp1-pwrc.usgs.gov/amphib/Frogcall.html.

RESOURCES

Brooks, R. P., D. A. Devlin, and J. Hassinger. 1993. *Wetlands and Wildlife.* Pennsylvania State University, University Park, PA.

Conant, R. 1975. *A Field Guide to Reptiles and Amphibians.* Houghton Mifflin Company, Boston, MA.

Firehock, K., L. Graff, J.V. Middleton, K.D. Starinchak, and C. Williams. 1998. *Save Our Streams: Handbook for Wetlands Conservation and Sustainability.* Izaak Walton League of America, Gaithersburg, MD.

Lingelbach, J. 1986. *Hands-On Nature.* Vermont Institute of Natural Science, Woodstock, VT.

National Wildlife Federation. 1998. *Ranger Rick's Nature Scope: Let's Hear It for Herps!* McGraw-Hill, New York, NY.

Smithberger, S. I. and C. W. Swarth. 1993. "Reptiles and Amphibians of the Jug Bay Wetlands Sanctuary," *The Maryland Naturalist.* 37(3-4):28-46.

White, C. P. 1989. *Chesapeake Bay: A Field Guide.* Tidewater Publishers, Centreville, MD.

Zim, H. S. and H. M. Smith. 1987. *Reptiles and Amphibians.* Golden Press, NY, NY.

On-line resources:

A Thousand Friends of Frogs; activities, information, teacher resources; cgee.hamline.edu/frogs/index.htm.

Amphibian Count Data Base; statistical analysis of available data; www.mp2-pwrc.usgs.gov/ampCV/ampdb.cfm.

Frogwatch USA; locate tapes of frog calls, local species lists, frog identification, pictures, range maps; www.mp2-pwrc.gov/frogwatch.

Frog Web; activities, monitoring programs, information; www.frogweb.gov.

Journey North Program; range maps, call recordings, shared data on frogs; www.learner.org/jnorth/jnorth.html.

North American Amphibian Monitoring Program; frog call survey, salamander monitoring, teachers' toolbox with experiments and data collection; www.mp1-pwrc.usgs.gov/amphibs.html.

Terrestrial Salamander Monitoring Program; salamander monitoring and experiments; www.im.nbs.gov/sally.

Spotted turtle,
Clemmys guttata

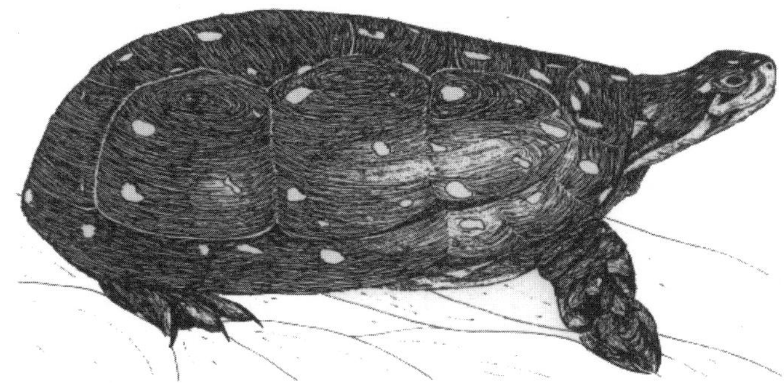

FROG SONG CARDS

You are a male Spring Peeper. Make the call: peep, peep, peep.	You are a female Spring Peeper. Listen for the call: peep, peep, peep.
You are a male Green Frog. Make the call: plunk, plunk, plunk.	You are a female Green Frog. Listen for the call: plunk, plunk, plunk.
You are a male Bullfrog. Make the call: ba-rum, ba-rum, ba-rum.	You are a female Bullfrog. Listen for the call: ba-rum, ba-rum, ba-rum.
You are a male Carpenter Frog. Make the call: pu-tunk, pu-tunk, pu-tunk.	You are a female Carpenter Frog. Listen for the call: pu-tunk, pu-tunk, pu-tunk.

RECORDS OF WETLAND HERPTILES

Date: ________________ Collection/Observation Method: ____________________________

Weather: ____

Time	Location	Species	Activity

Bugs and Butterflies

SUMMARY

Insect larvae and adults are preyed upon by larger wetland creatures. Consequently, many have developed strategies and adaptations that allow them to survive by not drawing attention to themselves. Butterflies and moths are beautiful exceptions. Several methods are presented to attract or find these often hidden creatures that are so important to the wetland food web, and whose presence can indicate wetland health.

OBJECTIVES

Students will sample adult and larval insects in the water, identify them by group, determine the proportion of each group, and evaluate water quality based on this data. Flying insects will be collected and identified. If species diversity is low, ways to attract additional beneficial species of insects to the wetland will be designed.

MATERIALS

- Film canisters containing scented objects
- Flashlights
- Field guides to insects and butterflies
- Butterfly nets
- Hand lenses or bug boxes
- D-nets
- White pans
- Coarse sieves
- Fine sieves
- Bottle trap and stake
- Copies of *Records of Wetland Insects* Student Page

MAKING CONNECTIONS

Bugs and butterflies can be lifelong sources of wonder. Butterflies and moths are beautiful additions to a flower garden, and are pollen carriers for many plant species. Beetles and other bugs come in many shapes, colors, and sizes. Their behavior often baffles us, but also amuses us.

Some species of insects indicate a healthy habitat, while others indicate a habitat that is out of balance. Some we appreciate because of their beauty; others we attempt to control because they bite or irritate us in some manner. Unfortunately, some people squash all insects without taking time to observe their unique characteristics or consider their role in the ecosystem.

BACKGROUND

To grow larger in size, insects shed their hard exoskeletons (molt) exposing new surfaces which expand before hardening into new exoskeletons.

Insects also change form as they grow, in a process called metamorphosis. The degree to which insects change as they grow differs among the major insect groups.

Some insects spend their entire lives in water. Some live in water during their early life stages, but emerge from the water for their adult stage. Still other insects live only on land.

With relatively brief life spans (usually one year or less) and typically short travel distances, insects are excellent indicators of water quality. Macroinvertebrates (mainly insects) have been used for many years in stream bioassessment, and are now being used in wetland bioassessment. Which assemblages (groups) work best for indicating wetland health is still being determined, but Ephemeroptera (mayflies), Odonata (dragonflies and damselflies), and Trichoptera (caddisflies) appear to be the most pollution-sensitive groups.

PROCEDURE

Warm-Up

With a small amount of preparation, this activity can be used year after year. Collect empty film canisters (available free at most photo shops) and place a small piece of scented soap, potpourri, or spice in each. Make sure there are at least two of each kind; mark some M for male and some F for female. Same-colored circles may be placed on the bottoms of containers with matching scents.

Pass out a scented film canister or flashlight (with tape giving the sex and code: short-short or short-long-short) to each student. Tell them they are insects looking for mates and can only reveal who they are by their scent or flashing light.

Female insects should stand still, but male insects will move about in search of a mate. Some "females" will give off a scent to attract a mate; "males" will sniff out a female that matches the scent in their canister. "Males" with flashlights will blink their pattern and watch for a female to answer with the same pattern. Only when a "female" sees a light pattern that matches hers will she reply, so the male can come to her. Each female may attract several males.

Have students remove the covers on their cannisters, and begin sniffing and flashing until they find partners. A partially darkened room provides an appropriate setting.

Activity

All grades

Try some of the following techniques to locate wetland insects. Keep personal and class lists of the insects sighted, the date, time of day, location, and weather on *Records of Wetland Insects* Student Page.

A. SIMPLE OBSERVATION (Individual or group)
Choose a single plant, such as a cattail. Examine it carefully. Try to identify all the creatures that use the plant as a habitat.

In the water, pick up a rock, stick, leaf, or other object. As the object begins to dry, insects on it will start to move. The objects may also be placed in a white pan with a little water. As the insects move around the bottom of the pan, they will be easier to see and identify. Record your findings, and release the creatures where they were discovered. Return any objects to their original locations, since they are also part of the habitat.

B. BUTTERFLY NET (Individual or group)
Use a light butterfly net with a deep bag to sweep through the tops of above-water portions of emergent vegetation, as well as through the herbaceous plants surrounding the wetland. After a sweep, turn the net handle to fold the bag over the rim and trap the insects inside.

If a hardy insect is captured, you can reach in and grab it with your fingers or scoop it into a bug box with a lid. If a butterfly or other fragile insect is captured, handle it very carefully to avoid injuring it. With the butterfly's wings up, grasp the butterfly gently near the base of the wings with thumb and forefinger. Once identified, gently release the insect where it was found. Record your findings.

C. D-NET SAMPLING (Individual or group)
While standing in shallow water, sweep the D-net from the surface to the bottom, then along the bottom a short distance, and back to the surface in one motion. The net is likely to be filled with vegetation and debris.

Spread the dipped material over a coarse screen or sieve placed above a white pan with water in it. Screen holes must be large enough for insects to crawl or fall through. Keep the screen or sieve over the pan of water for a period of ten minutes while the material is continually respread. Creatures will crawl out of the leaves and muck, and drop into the pan of water. After ten minutes, pour the pan of water through a fine sieve and return the creatures collected on the sieve to the pan with clean water. Identify the creatures by group, record your findings, and release the captured creatures where they were collected.

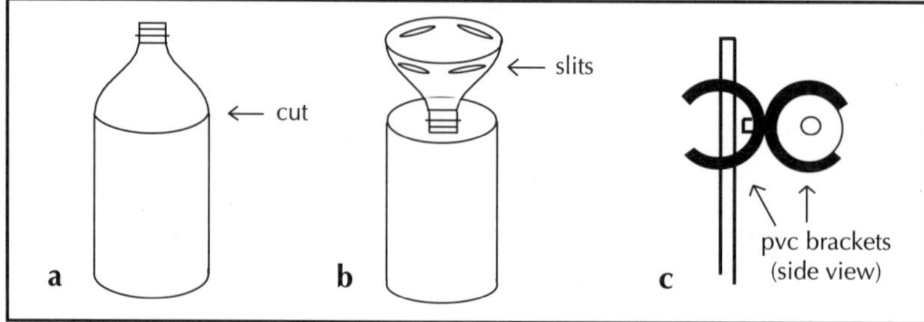

Figure 11.3 A bottle trap for collecting wetland insects can be fashioned from a plastic bottle that (a) is cut so that the top becomes a funnel, (b) has slits burned into the funnel portion to hold the funnel in place, and (c) is mounted horizontally on a rod with a bracket made from PVC pipe.

D. BOTTLE TRAP (Individual or small group)
Some predatory insects like beetles are usually missed when sampling with a D-net because they are fast swimmers or more active at night. A

funnel trap can be made from a two-liter, clear plastic beverage container attached to a stake. Leave the trap submerged in the wetland overnight. Creatures that swim into the trap will not be able to find their way out. Captured insects should be removed from the trap, identified, counted, and released. Be sure to record the information collected.

If you choose to build the trap, keep in mind that the entire trap should be as invisible as possible; yet capable of being opened, emptied of trapped insects, and reset. Cut the top from a two-liter plastic bottle at a point below the inward curve. When inverted into the bottom portion of the bottle, this top portion becomes the funnel. If four slits are cut into the wide bottom portion of the funnel with a hot tool (CAUTION!), the plastic will curl slightly so that it can be snapped into place. **Figure 11.3** provides some visual guidance. If this seems like a lot of work, the traps can be ordered (see Resources).

Grades 9-12

E. BIOASSESSMENT (Individual or group)

Bioassessment in wetlands usually involves collecting and identifying insects in the water, using techniques such as C and D above. Wetland bioassessment methods are not yet standardized, but are adapted stream bioassessment techniques that must be further modified for specific regional types of wetlands. Check with your state department of the environment and local nature centers to determine appropriate criteria for assessing local wetlands.

Pollution-sensitive wetland insects include the Orders Ephemeroptera (mayflies), Odonata (dragonflies and damselflies), and Trichoptera (caddisflies). Accurate identification of insects to family level is necessary for valid bioassessment and should be practiced before fieldwork begins.

General assessment tips:

- The greater the number of families in each of the three orders listed above, the more diverse and healthy the wetland is likely to be.
- The greater the degree of similarity of organisms between a wetland study site and an undisturbed reference site, the healthier the study site.
- A high ratio of pollution-tolerant to pollution-intolerant organisms indicates a less healthy wetland.

Bioassessment may include more than insects. See *Aquatic Animals*, Macroinvertebrate Bioassessment on page 307.

Wrap-Up

Compile the data collected from all sources and record it on the board, an overhead transparency, or a poster. What were the three most abundant groups of insects collected in the water? Are these the three pollution-sensitive groups?

What are the most abundant flying insects? Can the wetland be enhanced with additional plants that will attract more butterflies?

ASSESSMENT

Did students successfully collect and identify insects? Have students do the following:

- Name and sketch the three most abundant groups of insects collected from the water, and (based on their observations) explain why they were the most abundant.
- Name, and sketch or describe the most abundant flying insects (butterflies, dragonflies, etc.) collected.
- Characterize the wetland habitat in which the insects were found.
- Explain what would attract more beneficial insects to the wetland.

EXTENSIONS

Determine the sources of wetland stress, if they exist. Can changes be made inside or outside the wetland that will improve the habitat for the insects that are so important to the food chain?

Enhance the wetland with plants that attract butterflies.

Join a local or regional volunteer group that conducts wetland bioassessments, or supply data collected by your group. Continue monitoring the wetland for changes in the insect populations, especially if land use within the watershed is changing.

RESOURCES

Borror, D. J. and R. E. White. 1970. *A Field Guide to the Insects.* Houghton Mifflin Company, Boston, MA.

Brooks, R. P., D. A. Devlin, and J. Hassinger. 1993. *Wetlands and Wildlife.* Pennsylvania State University, University Park, PA.

Cornell, J. 1989. *Sharing the Joy of Nature.* Dawn Publications, Nevada City, CA.

Danielson, T. J. 1998. Wetland Biological Assessment Fact Sheets. USEPA, Office of Wetlands, Oceans, and Watersheds, Washington, DC (EPA/843/F/98/001).

Ely, E., Editor. Spring 1998. Multiple articles on monitoring wetlands. *The Volunteer Monitor*. Vol. 10, No. 1. San Francisco, CA.

Fadala, S. 1989. *Basic Projects in Wildlife Watching.* Stackpole Books, Harrisburg, PA.

Hicks, A. 1997. *Freshwater Wetlands Invertebrate Biomonitoring Protocol for the New England Region.* University of Massachusetts, Amherst, MA.

Kellogg, L. L. 1992. *Monitor's Guide to Aquatic Macroinvertebrates.* Izaak Walton League of America, Gaithersburg, MD.

Lingelbach, J. 1986. *Hands-On Nature: Information and Activities for Exploring the Environment with Children.* Vermont Institute of Natural Science, Woodstock, VT.

McCafferty, W. P. 1983. *Aquatic Entomology*. Jones and Bartlett Publishers, Inc., Sudbury, MA.

Merrit, R. W. and K. W. Cummins (editors). 1984. *An Introduction to the Aquatic Insects of North America.* Kendall/Hunt Publishing Company, Debuque, IA.

Niering, W. A. 1992. *The Audubon Society Nature Guides—Wetlands.* Alfred A. Knopf, Inc., New York, NY.

Reid, G. K. 1987. *Pond Life*. Golden Press, New York, NY.
Woodbury, E. N. 1994. *Butterflies of Delmarva*. Tidewater Publishers, Centreville, MD.
Zim, H. S. and C. Cottam. 1987. *Insects*. Golden Press, New York, NY.

To order the bottle trap, contact Don Wik at E4596 266th Avenue, Menomonie, WI 54751.

On-line resources:
Butterfly Monitoring; www.mp1-pwrc.usgs.gov/butterflies.html.
Journey North Program; monarch butterfly migrations; www.learner.org/jnorth/jnorth.html.
North American Butterfly Association; www.naba.org.
Save Our Streams macroinvertebrate key; www.people.Virginia.EDU/~sos-iwla/Stream-Study/StreamStudyHomePage/StreamStudy.HTML.
U.S. EPA Wetlands Division Website; indicators, techniques, and applications of community level biomonitoring data; www.epa.gov/owow/wetlands/wqual.introweb.html.

Viceroy butterfly,
Basilarchia archippus

RECORDS OF WETLAND INSECTS

Date: ________________ Collection/Observation Method: ____________________________

Weather: ____

Time	Location	Species	Activity

Aquatic Animals

SUMMARY

A freshwater wetland is as productive as a rainforest, while a salt marsh is even more productive, acre for acre, than a tropical rainforest. This is due to the wetlands' functions as a producer of vegetation, a recycler of detritus, and a nursery for various creatures. A large variety of fish, crustaceans, and mollusks inhabit a wetland during all or part of their life cycles. The diversity and relative abundance of aquatic animals will be investigated.

OBJECTIVES

Fish, crustaceans, and mollusks found in wetlands will be captured, identified, examined, and their abundances compared.

MATERIALS

- Dip nets
- D-nets
- Seine net
- Life jackets
- Minnow trap
- Buckets
- White pans
- Egg cartons (not cardboard)
- Field guides
- Copies of *Records of Aquatic Animals* Student Page

MAKING CONNECTIONS

Saltwater and freshwater wetlands are major sources of food for our growing world population. Oysters, clams, mussels, crabs, crayfish, and fish are dependent on wetlands for all or part of their life cycles. Clams, mussels, and oysters live in shallow water that produces an abundance of algae on which they feed. Crabs spend most of their lives in wetlands, hiding in grass beds where they can most safely shed their exoskeletons when they become too small, and preying on smaller creatures found there. Many fish species use wetlands as a nursery during their larval and juvenile stages, again because of the abundance of food and hiding places. Fish that are too large as adults to live in wetlands feed on smaller fish that do live there. So much of our seafood depends on wetlands, that without wetlands our supply of seafood would dwindle.

BACKGROUND

Aquatic animals come in a variety of forms and sizes. Some live only in saltwater, some only in freshwater, and some only in brackish waters. A few animals have the unique ability to migrate between the extremes of salinity. All aquatic animals have optimal ranges for the physical parameters of their habitat. For example, rainbow trout favor cold, fast-moving, freshwater; they cannot be found in warm ponds or ocean water. Lobsters are found on the ocean bottom, while their crayfish cousins are found in freshwater streams, ponds, and lakes. Knowing what conditions are favored helps us to find these creatures.

Aquatic animals can be collected actively or passively, with expensive equipment or simple devices. The key is having a wetland to investigate and knowing where to search.

PROCEDURE

Warm-Up

Grades K-4

Match the creature with the appropriate characteristic:

Crab	Two hinged shells, attaches to something hard, saltwater
Minnow	A tail, a shell that must be shed to grow, freshwater
Clam	Two claws, a hard shell, no tail
Oyster	A tail and scales, eaten by many creatures
Crayfish	Two hinged shells, not attached to anything

Grades 5-12

Have students describe the characteristics and habits of the creatures listed above.

Activity

All grades

Most of the collecting techniques presented are appropriate for all grades. Depending on the age and skills of the students participating, **life jackets may be appropriate**.

Identification of the creatures collected, the date, time of day, location, and weather should be kept on the *Records of Aquatic Animals* Student Page inside a classroom Wetland Log Book. This will allow comparison of data for different tides, seasons, and years. Changes to the wetland over time will thus be documented and can be correlated to changes in land use within the watershed.

A. MINNOW TRAPS (Individual or group activity)
An unbaited minnow trap is one of the easiest sampling devices to use in a wetland. Creatures enter the minnow trap because it has structure and, therefore, offers safety from predators, not because it is baited with food. Small fish, shrimp, crabs, and nettles can be collected in salt or brackish waters along docks, bulkheads, or other structures to which a trap can be secured by a line. In freshwater wetlands, secure the minnow trap to a tree or stake along the bank, making sure the trap is not completely submerged so that small turtles that wander in do not drown. Check the trap daily and record the number and variety of species collected. Be sure to release the captured creatures.

B. HARD STRUCTURE (Small group activity)
From a dock or bulkhead, suspend a mesh bag of old shells, a piece of cinder block, or whatever else is at hand that is hard enough for organisms to attach. Be sure that the object hangs low enough to be covered with water most of the time.

Once a month, pull up the bag or cinder block and place it in a large white pan or bucket. Pour a little water on it and watch creatures venture from their hiding places. Look carefully to see which creatures remain attached to the hard surfaces (oyster spat, mussels, anemones, sea squirts, corals, sponges). After listing the species that have colonized the new habitat, carefully return them to the water along with the hard structure. Keep a record of which species are present, their size, and the date. How long has it taken each species to colonize the habitat? Try putting out another object each month, and compare colonization rates at different times of the year.

If the wetland is shallow, try the same activity with a rock, cinder block, or a mesh bag of smaller rocks placed directly on the bottom in an accessible area. Creatures will eventually attach to the hard structure or hide in, around, and under it.

C. DIP NETS AND D-NETS (Group activity)
Dip nets can be as small as those used in aquaria, or have long handles for searching under stream banks and in tide pools. If one side of the frame supporting the net is flat, it is called a D-net. The D-net is most effective for collecting bottom-dwelling creatures. The task of collecting creatures can be made difficult by the presence of vegetation, but that is where creatures hide. If vegetation gets tangled in the nets, or algae clogs them, the nets are designed to be turned over and cleared by reversing the water flow. Check first, however, to see what creatures have been caught.

White buckets, white pans, ice cube trays, and egg cartons are useful for observing the creatures collected. Handle the creatures as little as possible. When handling them, make sure hands are wet to prevent damage to the protective slime coating of the creatures. Use a field guide to determine the species caught. Keep records of creatures by species, size (measure fish from tip of nose to tip of tail), date, time, and place.

D. EDGE SEARCH (Group activity)
An easy collecting technique is to walk along the water's edge, watching for creatures. If the wetland is tidal, low tide will be the best time for exploring the exposed sand or mud flats. Whelks, horseshoe crabs, fiddler crabs, snails, and hermit crabs are easily collected in this manner in saltwater wetlands. Watch for whelk and skate egg cases, horseshoe crab eggs, and the sand collars produced when moon snails lay their eggs. In freshwater marshes, look for crayfish and the chimneys that mark their burrows. Examine creatures gently, then be sure to return them to the locations where they were found. Record the date, time, tide level (if it's a tidal wetland), species, and activities of animals.

E. BOTTOM SAMPLING (Small group activity)
A bottom grab sampler may be purchased for use from a small bridge, pier, walkway, or bulkhead. If the water is shallow, students may stand in the water and scoop up some bottom material with a cup, plastic container, or large spoon. Empty the bottom material into a shallow white pan with some water. Wriggling creatures may be fished out with fingers or forceps, and placed in water held in compartmentalized containers (such as egg cartons or ice cube trays) for examination and identification. Record all information on a data sheet.

Grades 5-12
Height and strength are limiting factors in seining. Younger students willing to team up and work with smaller seine nets may be able to seine.

F. SEINING (Group activity)
In shallow freshwater wetlands, short four-foot seines can be used. Along mud flats and sandy beaches associated with salt marshes, larger nets that are ten to fifty feet in length may be used, depending on the size and strength of the students handling them. Seining is a collection technique that requires several people in boots (if the water is shallow), waders (if deep and cold), or swimsuits (if deep and warm). Simple seine nets with two sturdy poles can be inexpensive and more fun than fancier models. These are great sampling devices for open water areas, and the creatures collected will be more numerous and larger than those collected by other techniques, provided good sampling techniques are employed.

In small wetlands with a current, a four-foot kick seine is useful for finding creatures. One or two people extend the kick seine across the water, making sure the bottom of the net stays on the bottom and that they are on the downstream side of the net. One or two other students move upstream of the net, shuffling their feet along the bottom. After the creatures have had a chance to drift into the net, lift the net out of the water. Gently collect the creatures with fingers or forceps. Place them in a pan or jar containing water from where they were collected, and examine the creatures. Be sure to return them to the water where they were collected within a short period of time. As water warms, it quickly loses dissolved oxygen; and the creatures could suffocate.

To sample a sandy beach or mud flat, use a ten-foot to fifty-foot seine. Begin on land with one person holding each pole (two per pole if students are small and the net large), and the seine net stretched loosely between them (weights down, floats up). Walk into the water, keeping the lower end of the poles and net angled forward and against the bottom. Before reaching the limits of water depth, the shorter person slows while the taller person sweeps around that student; both then proceed back to land. As land is approached, **the lower edge of the net must be kept on the bottom until the net is on shore!** If the net is lifted prematurely, all or most of the creatures in it will escape under the net. Once the lower edge of the net is on shore, the two people holding the poles step apart, lifting the bottom edge of the net. They then lay the net flat on the beach or mud flat. When the seine net is large or heavy with creatures and debris, a third person may step in to help lift and fold the middle of the net closed to prevent loss of the netted creatures and damage to the net.

With a handy bucket of water in which to place them, the captured creatures can be removed from the net quickly and with little stress. Check balled up pieces of vegetation for hidden creatures. Watch for wriggling motions since some creatures are translucent (such as shrimp). Make sure hands are wet when handling creatures. Also, watch out for sharp spines, claws, and teeth. If needed, have someone with experience demonstrate safe handling of the creatures. Record the number of species collected, the date, time, location, and water depth. Return the creatures to the water where they were collected as quickly as possible.

G. MACROINVERTEBRATE BIOASSESSMENT (Individual or group)
Macroinvertebrates are animals without backbones that are large enough to see without magnification. Insects are the major group used in bioassessment, but worms, snails, and clams are also indicators of water quality. The presence of gilled (right-handed) snails indicates good water quality, because they are very sensitive to pollutants and low oxygen levels. Crayfish and clams are less sensitive. Lunged (left-handed) snails, planaria, and leeches are tolerant of pollution, but can also be found in good quality water.

As water becomes polluted, two things happen. Species richness declines and the population density of the remaining species increases. In other words, there will be fewer species, but many creatures of each remaining kind. Check with the Department of the Environment in your state to determine what species are used as indicator species to assess wetlands in your area.

Using the bottom sampling technique in section E above, collect benthic macroinvertebrates. Identify the organisms with an appropriate key. Record the species names, number collected, location, weather, and date. Instead of activity, list the pollution tolerance level. If information on regional bioassessment standards is not readily available, compare species richness and population numbers with other local wetlands of the same type to determine relative levels of pollution.

Wrap-Up

Carefully tally the data collected. Record it on the blackboard, a poster, or overhead transparency so all students can see it. How many species were found? Which species were most abundant? Which was rarest? Why are these species found in the wetland? Which species stay in the wetland their entire life?

ASSESSMENT

Based on the data collected, have students do the following:

- Rank the species from most abundant to least abundant, or construct graphs showing the relative abundance of each species found in the wetland.
- Explain why some species are more abundant than others.
- If some species that students expected to find are missing from the wetland, determine possible reasons for their absence.

EXTENSIONS

Through research, determine what each of the collected creatures eats and what eats them. Construct a food web showing the feeding relationships. What would happen if one member of the food web was eliminated? What if three members of the food web were eliminated? Investigate energy pyramids and construct one.

RESOURCES

Amos, W. H. and S. H. Amos. 1989. *The Audubon Society Nature Guides: Atlantic and Gulf Coasts.* Alfred A. Knopf, New York, NY.

Caduto, M. J. 1990. *Pond and Brook: A Guide to Freshwater Environments.* University Press of New England, Hanover, NH.

Coulombe, D. A. 1987. *The Seaside Naturalist.* Prentice Hall Press, New York, NY.

Gosner, K. L. 1979. *The Peterson Field Guide Series: A Field Guide to the Atlantic Seashore.* Houghton Mifflin, Boston, MA.

Kesselheim, A. S. and B. E. Slattery. 1995. *WOW!: The Wonders of Wetlands.* Environmental Concern Inc., St. Michaels, MD.

Lippson, A. J. and R. L. Lippson, 1997. *Life in the Chesapeake Bay.* The Johns Hopkins University Press, Baltimore, MD.

Niering, W. A. 1985. *The Audubon Society Nature Guides: Wetlands.* Alfred A. Knopf, Inc., New York, NY.

Pennak, R. W. 1989. *Freshwater Invertebrates of the United States.* John Wiley and Sons Inc., Somerset, NJ.

Reid, G. 1987. *A Golden Guide: Pond Life.* Golden Press, New York, NY.

Ripple, K. L. 1998. Wetland Creatures in Saltwater Aquaria. *Wetland Journal* 10(2):3-7.

Robins, C. R. and G. C. Ray. 1986. *The Peterson Field Guide Series: A Field Guide to Atlantic Coast Fishes of North America.* Houghton Mifflin Company, Boston, MA.

White, C. P. 1989. *Chesapeake Bay: A Field Guide.* Tidewater Publishers, Centreville, MD.

Zim, H. S. and H. H. Shoemaker. 1987. *A Golden Guide: Fishes.* Golden Press, New York, NY.

On-line resources:

Hooked for Life; a comprehensive fishing site for students of all ages; www.WorldWideAngler.com/hookedforlife.

Save Our Streams macroinvertebrate key; www.people.Virginia.EDU/~sos-iwla/Stream-Study/StreamStudyHomePage/StreamStudy.HTML

U.S. EPA Wetlands Division Website; indicators, techniques, and applications of community level biomonitoring data; www.epa.gov/owow/wetlands/wqual.introweb.html.

U.S. Fish and Wildlife Service; information on fish; www.fws.gov.

RECORDS OF AQUATIC ANIMALS

Date: ________________ Collection/Observation Method: __________________________

Weather: ____

Time	Location	Species	Activity

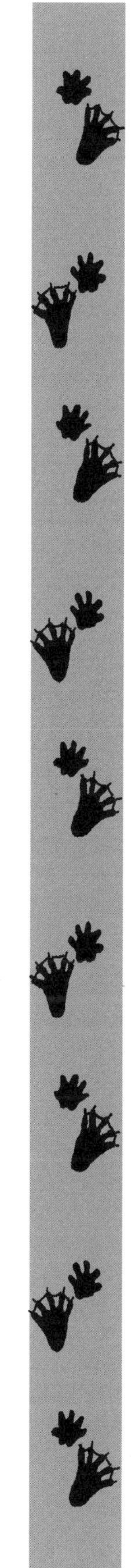

Wetland Soup

SUMMARY

Microorganisms play an important role in wetland food webs. The types of organisms, their abundance, and the timing of their presence will be investigated.

OBJECTIVES

The microorganisms present in a wetland will be sampled, counted, and classified by category. Their relative abundances throughout the year will be charted, analyzed, and compared with abundances of other wetland creatures to determine interrelationships.

MATERIALS

- Field guide to microorganisms
- Microscopes (or projecting microscope)
- Slides
- Eyedroppers or pipettes
- Chemical to slow microorganism activity
- Sample containers
- Plankton net with line attached
- Fishing weight
- Water sampler with line attached
- Bottom grab sampler with line attached
- Disposable bacterial samplers for *E. coli*
- Copies of *Records of Wetland Microorganisms* Student Page

MAKING CONNECTIONS

Wetlands produce more food per acre than most other ecosystems, with tidal salt marshes surpassing even rainforests in productivity. This amazingly high level of productivity is achieved in wetlands by both the growth of plants and the decomposition of detritus by bacteria forming a rich wetland soup. Microscopic herbivores and detritivores are consumed by zooplankton, which are then eaten by larger aquatic creatures. Most aquatic organisms consumed by humans begin life as larvae feeding on the rich soup of wetlands.

BACKGROUND

All living organisms have a food source. Producers such as algae make their own food in a process called photosynthesis. Consumers (copepods, larvae of crabs and fish, etc.) eat those who produce the food. Decomposers (bacteria and fungi) obtain their nourishment by breaking down dead organic matter called detritus–a major food source in wetlands. Nematode worms and insect larvae are some of the many benthic organisms that feed on the decomposing detritus. The list of ingredients of wetland soup includes members of all these categories.

Some organisms (like protozoans) will remain small all their lives; others (such as fish larvae) may become quite large. Some spend their entire lives in the wetland; others may migrate out of the wetland as they develop into adults, returning later to lay their eggs.

PROCEDURE

Warm-Up

Which would be easier to eat: a whole watermelon (uncut and not smashed) or a slice of watermelon? An entire cow or a burger? Why? Humans, like most organisms, are adapted to eating foods that are broken down into smaller units; we have invented knives to help us do that. In the natural world many microorganisms are adapted to breaking down large aquatic plants and animals into smaller, less complex units that other organisms can utilize.

Activity

All grades

Microscopes will be needed to see the microorganisms. In classes of students too young to handle microscopes on their own, a projecting microscope controlled by the teacher is a useful alternative. Try some or all of the following activities. Be sure to save data and observations on *Records of Wetland Microorganisms* Student Page.

A. PLANKTON SAMPLING (Small or large groups)
To make a plankton net, obtain some fine mesh material such as knee-high stockings. Attach the open end of each stocking to an embroidery hoop of similar size.

Using either the homemade plankton net or a purchased plankton net, attach three lines, each about two feet long, to the embroidery hoop at three points. Connect the three lines to another longer line to form a three-point harness and towline. See **Figure 11.4**.

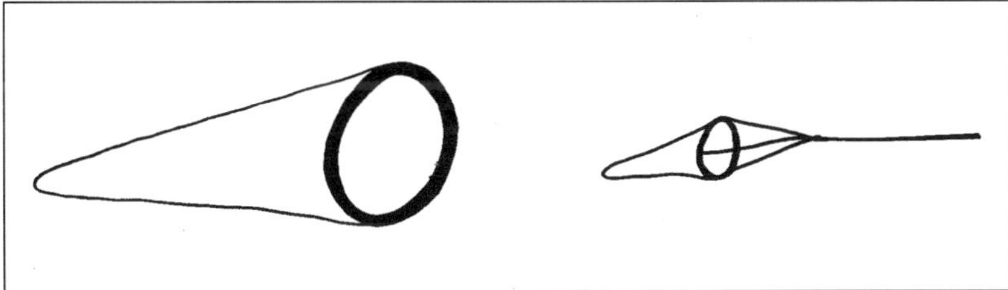

Figure 11.4 A plankton net needs a three-point harness and towline. A plastic or glass sample bottle may be placed inside the cod (small) end of the net and secured with a heavy rubber band.

Walk along the edge or across the middle of a wetland while towing the plankton net at the surface. Remind students that they are collecting *micro*organisms, not things that they can see! Invert the net into a sample jar of water and gently shake the organisms into the water for examination with a microscope. Place the sample jar of water and organisms in the shade or in a cooler for transport back to the classroom for analysis.

Attach a small fishing weight to one edge of the net and repeat the earlier tow with the plankton net several inches below the surface of the water. Again invert the net into a sample jar of water, gently shake the organisms loose, and store them in the shade or a cooler for transport.

IN THE CLASSROOM:
From the sample jar of water containing organisms from the surface, take a sample from the water surface using the pipette or eyedropper. Place it on a slide (flat or with a well), cover with a cover slip, and examine it using a microscope. If the microorganisms are moving too fast to be easily observed, add a drop of chemical to slow them down. Keep in mind that when the chemical is added all microorganisms will slow down. It will be more difficult to recognize the living organisms when they do not move.

Within one field of view:

- How many organisms appear to have some green color? (algae)
- How many appear to be slowly spinning around whether or not they are attached to the slide? (rotifers)
- How many propel themselves along a spiral path? (dinoflagellates)
- How many move quickly in a jerking, almost leaping fashion? (copepods)
- How many move like worms across the slide? (nematodes and other worms)
- Do any of the organisms appear to have shells like clams or snails? (clam, snail, and oyster larvae)
- Do any swim like fish? (fish larvae)
- Are there any round eggs on the slide? Can you see the eyes or heart of an organism inside an egg?

Try to draw some of the microorganisms or sketch their motions.

Repeat by withdrawing a pipette of material from the middle of the sample jar of water; then repeat with one from the bottom of the jar. Repeat with the sample jar of water containing organisms collected below the surface. Are there similarities among the samples? How are they different? Which group of microorganisms is most abundant in each sample?

B. DEEP WATER SAMPLING (Small or large groups)
Water sampling bottles can be purchased or constructed. A water sampler can be made of a glass bottle fitted with a cork to close it and a fishing weight to take it down to the desired depth. Attach a sturdy string to the neck of the bottle and another string into an eyelet screwed into the cork.

Lower the bottle to the desired depth (surface, midwater, or bottom), and pull the cork from the bottle. When bubbles stop coming to the water surface, lift the bottle straight up without tipping it over. Pour the water sample into another clean sample bottle, and place it in the shade or in a cooler for later analysis. Take water samples at the other two levels. Follow the same procedure for classroom analysis as in the previous activity (A).

C. BOTTOM SAMPLING (Small or large groups)
If the water is shallow enough to reach the bottom by hand, use a plastic container or large spoon to scoop up bottom material. If the water is too

deep to reach by hand, a small bottom grab sampler may be purchased. Place the sample of bottom material in a sample container, add water to keep it moist, and place the sample in the shade or a cooler for transport back to the classroom. Follow the same procedure for classroom analysis as in the first activity (A).

Most decomposition occurring in the bottom muck is anaerobic (without oxygen). What does the sample smell like? What does the sample feel like?

Grades 9-12

D. FECAL COLIFORM SAMPLING (Individual or small group)
Fecal coliform bacteria (*Escherichia coli* or *E. coli* for short) are indicators of pollution by human waste products. If sewage pollution of the wetland is suspected, no hands-on activities should be engaged in until the water has been tested. An easy method is to use disposable samplers available in many biological supply catalogs. While wearing rubber gloves, open the plastic sampler and dip it in the water (follow manufacturers' directions). Close the sampler and incubate at the temperature specified (usually a closed classroom cabinet is fine). Follow directions for analyzing the results. Other than the sampler, no special equipment is needed. If results are positive, contact the local health department for further testing. Cease all hands-on activities in the wetland.

Wrap-Up

On an overhead transparency or poster, summarize the class data on the abundance of each group of microorganisms. Was one group found most often (or only) in surface samples? If so, which one? Which group of microorganisms was found only in bottom samples? Why? Were any groups of microorganisms found only in midwater samples? Which ones? What might be the cause of this? Record the data on the abundance of each group of microorganisms at each water level. Repeat the activity every month (or each season) to determine how relative abundances change throughout the year.

ASSESSMENT

- Did students use good sampling techniques?
- Were microscopes used carefully and appropriately?
- Are student drawings and/or sketches of motions reasonable representations of the organisms seen?
- Are student data summaries complete and accurate?
- Can students describe the patterns of abundance within the water column and give possible reasons for this pattern?
- Can students explain interrelationships between wetland microorganisms and other wetland creatures. Are the interrelationships supported by the data?

EXTENSION

Compare data on microorganisms with data on other wetland creatures. Do abundances of microorganisms correspond to population changes in other groups of wetland creatures? Can interrelationships be deduced from this? How could these ideas be tested?

Continue sampling each month throughout the year. Make graphs of the relative abundances of each group of microorganisms each month. Compare these populations with the abundances of other animal groups, such as fish or frogs. Are there any correlations? Why or why not?

RESOURCES

Caduto, M. J. 1990. *Pond and Brook.* University Press of New England, Hanover, NH.

Coulombe, D. E. 1987. *The Seaside Naturalist.* Prentice Hall Press, New York, NY.

Lingelbach, J. 1986. *Hands-On Nature: Information and Activities for Exploring the Environment with Children.* Vermont Institute of Natural Science, Woodstock, VT.

Lippson, A. J. and R. L. Lippson. 1997. *Life in the Chesapeake Bay,* Second Edition. The Johns Hopkins University Press, Baltimore, MD.

Reid, G. 1987. *Pond Life.* Golden Press. New York, NY.

Mitsch, W. J. and J. G. Gosselink. 1993. *Wetlands,* Second Edition. Von Nostrand Reinhold, New York, NY.

White, C. P. 1989. *Chesapeake Bay: A Field Guide.* Tidewater Publishers, Centreville, MD.

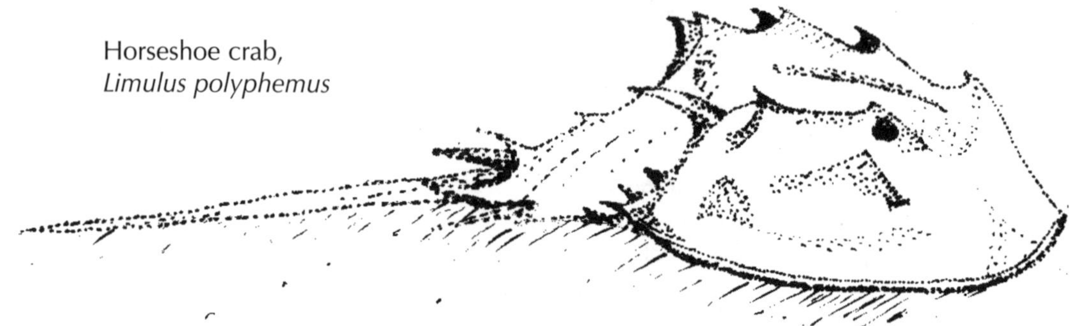

Horseshoe crab,
Limulus polyphemus

RECORDS OF WETLAND MICROORGANISMS

Date: ____________ Time: __________ Collection/Observation Method: ______________________________

Weather: ____

Sample Number	Algae	Copepods	Worms	Rotifers	Dino-flagellates	Mollusk Larvae	Fish Larvae	Other

12 WETLAND STEWARDSHIP

International Stewardship

National Stewardship

State and Regional Stewardship

Schoolyard Habitats Application

International Stewardship

Wetlands all over the world are in need of protection as we strive to undo the damage done in the past and learn to support these imperiled ecosystems.

THE RAMSAR CONVENTION ON WETLANDS

An intergovernmental treaty to conserve wetlands, The Convention on Wetlands was first signed on February 2, 1971, in the Iranian city of Ramsar. As of January 2000, the Ramsar Convention had been signed by 119 countries; and 1,021 wetland sites within those countries had been listed as Wetlands of International Importance. These wetlands range in size from a one-hectare mangrove ecosystem in Australia to a 6.8 million-hectare delta in Botswana. In location they range from the Arctic Circle in Canada to an Atlantic Coast Reserve in Argentina.

The vision for the Ramsar List of Wetlands of International Importance is "to develop and maintain an international network of wetlands which are important for the conservation of global biological diversity and for sustaining human life through the ecological and hydrological functions they perform." Member countries that have signed the Ramsar Convention are obliged to determine which of their wetlands should be considered for the Ramsar List, then plan how those wetlands will be preserved in good health. Assistance is provided by the Ramsar Convention Bureau in Gland, Switzerland and through technical experts, training sessions, publications, videos, and a web site.

To access the Ramsar List of Wetlands of International Importance for specific countries, log on to the Ramsar web site (www.ramsar.org), click on Country Profiles, then click on the country of choice.

As of January 2000, the United States has 17 wetlands included on the Ramsar List. If a site is nearby, a field trip there may result in sensory experiences and research comparisons not yet available in a schoolyard wetland. Wetlands on the Ramsar List are recognized and protected, but they should also be celebrated as the unique and important treasures that they are.

WORLD WETLANDS DAY

February 2nd is recognized as World Wetlands Day, providing a context for celebrating wetlands of international importance in countries around the world. Posters, buttons, and other educational materials are available in several languages from the Ramsar Bureau in Switzerland. Groups may also post their plans and/or the results of their World Wetlands Day celebrations on the Ramsar web site (www.ramsar.org). This web site also provides wetland information for many educational levels, from elementary to technical, and in several languages (primarily English, Spanish, and French).

National Stewardship

Several national wetland stewardship programs established by government agencies and nonprofit organizations are suitable for school groups and students.

AMERICAN WETLANDS MONTH

In 1991 the Terrene Institute launched the celebration of May as American Wetlands Month to promote wetlands awareness, education, and action using posters, publications, videos, and conferences. Beginning in 2000, the Save Our Streams Program of the Izaak Walton League of America assumed sponsorship of American Wetlands Month events and the Communities Working for Wetlands conferences. Check both web sites (www.terrene.org and www.iwla.org) for wetlands information or contact them directly at Terrene Institute, 4 Herbert Street, Alexandria, VA 22305 or Save Our Streams Program at 1-800-BUG-IWLA.

SCHOOLYARD HABITATS PROGRAM

While not devoted exclusively to wetlands, the Schoolyard Habitats Program of the National Wildlife Federation, includes wetland habitats within its certification program. A created, restored, or enhanced wetland on or near school grounds that is part of an educational program may be certified and entered in the National Register of Schoolyard Habitats. Upon certification, the school receives a Certificate of Achievement and a subscription to the program newsletter. An announcement of the achievement is sent by the National Wildlife Federation to local news media to further celebrate the event.

An Application for Certification is provided on pages 323-326. It may be completed by a class or a school and submitted to the National Wildlife Federation, Schoolyard Habitats Program, 8925 Leesburg Pike, Vienna, VA 22184-0001. A Schoolyard Habitats Information Kit containing additional information, contacts, and support materials is available for a small fee at www.nwf.org/habitats/schoolyard/index.html.

Home schooling programs might prefer to submit an application for recognition within the Backyard Habitats program. Additional information about this program is available at www.nwf.org/habitats/backyard.

U. S. ENVIRONMENTAL PROTECTION AGENCY

The U.S. EPA provides wetland information, materials, and links, including a wetlands information kit and instructions on how to adopt a wetland. Contact the Public Information Center (PM-211B), U.S. EPA, 401 M Street SW, Washington, DC 20460; telephone 1-800-832-7828; or e-mail the Wetlands-Hotline@epa.gov. A tremendous amount of material is available at the web site: www.epa.gov/owow/wetlands.

The Water Drop Patch developed by the U.S. EPA in conjunction with the Girl Scout Council of the Nation's Capital encourages Girl Scouts to "make a difference in their communities by becoming watershed and wetlands stewards". For more information, contact your local Girl Scout Council or visit www.epa.gov/adopt/patch/index.html.

U. S. FISH AND WILDLIFE SERVICE

Regional Wetlands Coordinators are available in all seven areas of the United States, providing information and services relating to the National Wetlands Inventory. NWI maps, information on wetland plants, wetland links, reports, publications, and other materials for educators can be accessed through the web site www.nwi.fws.gov. Contact information for each regional coordinator is provided as well.

Swamp rose mallow,
Hibiscus mocheutos

State and Regional Stewardship

Governmental groups across the continent recognize that concerned citizens and students of all ages need to be involved in the protection of wetlands if wetland loss and degradation are to be effectively halted. Legislators have passed laws that government agencies apply and enforce. Citizens, however, need to participate in a positive way in order for the protection process to be successful. Appreciating the beauty and wonder of wetlands and understanding the functions and values of wetlands are not enough. Many of us want to be personally involved in protecting and celebrating our treasured wetlands.

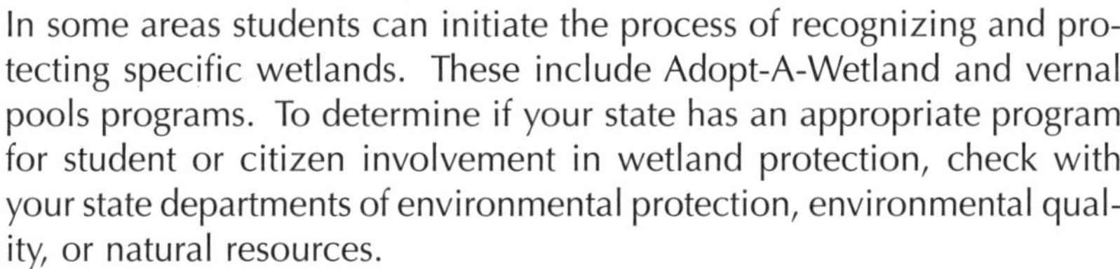

In some areas students can initiate the process of recognizing and protecting specific wetlands. These include Adopt-A-Wetland and vernal pools programs. To determine if your state has an appropriate program for student or citizen involvement in wetland protection, check with your state departments of environmental protection, environmental quality, or natural resources.

Following are summaries of some regional wetland programs.

GREAT LAKES

The Great Lakes Marsh Monitoring Program focuses on locating and identifying amphibians and marsh birds, then describing the habitat along monitoring routes surrounding all of the Great Lakes. For a detailed description of the program, visit the web site www.bsc-eoc.org/mmpmain.html. To become volunteers contact Kathy Jones, Aquatic Surveys Officer, The Marsh Monitoring Program, Long Point Bird Observatory, P. O. Box 160, Port Rowan, Ontario Canada N0E 1M0; telephone 519-586-3531; e-mail aqsurvey@bsc-eoc.org.

GULF OF MEXICO

Girl Scouts in the states of Alabama, Florida, Louisiana, Mississippi, and Texas are eligible to earn a Gulf of Mexico Patch upon completion of a variety of wetland and watershed activities that includes adoption of a wetland. For further information, contact a local Girl Scout Council.

NEW ENGLAND

Schools within US EPA Region 1 (Connecticut, Maine, Massachusetts, New Hampshire, Rhode Island, and Vermont) are eligible for a regional Adopt-A-Wetland Program. Any class that participates in one or more wetland stewardship activities, then completes an application with documentation of the activities may receive an Adopt-A-Wetland Certificate. For information and application forms, contact EPA Region 1, Wetland Protection Section, JFK Federal Building, Boston, MA 02203, Attention: Adopt-A-Wetland Program; telephone 617-565-4868 or access massbay.mit.edu/citizen.html.

A certification application form and stewardship activities are part of the wetlands curriculum, *A World in Our Backyard,* which was produced by the EPA. The guidebook and video are available for a fee through Environmental Media Corporation, P. O. Box 1016, Chapel Hill, NC 27514; telephone 1-800-368-3382.

REGION 4, U. S. ENVIRONMENTAL PROTECTION AGENCY

Region 4 of the U. S. Environmental Protection Agency (Alabama, Florida, Georgia, Kentucky, Mississippi, North Carolina, South Carolina, Tennessee) provides information and assistance to class, club, or community groups interested in protecting wetlands through a regional Adopt-A-Wetland project. Contact Gail Williams-Harrison or Mickey Feltus at US EPA Region 4, Wetlands Protection Section, 61 Forsyth Street SW, Atlanta, GA 30303-8960; web site www.epa.gov/region4/water/wetlands/education/adoptb.html; telephone 404-562-9410.

REGION 8, U. S. ENVIRONMENTAL PROTECTION AGENCY

The U. S. Environmental Protection Agency has taken the lead in Adopt-A-Wetland programs within Region 8 (Colorado, Montana, North Dakota, South Dakota, Utah, Wyoming), placing emphasis on protection of wetland integrity while maintaining diverse habitat and other values. Contact the Water Quality Requirements Section, One Denver Place, Suite 500, 999 18th Street, Denver, CO 80202-2405 or telephone 303-293-1570. Check the web site at www.epa.gov/region08 for current information.

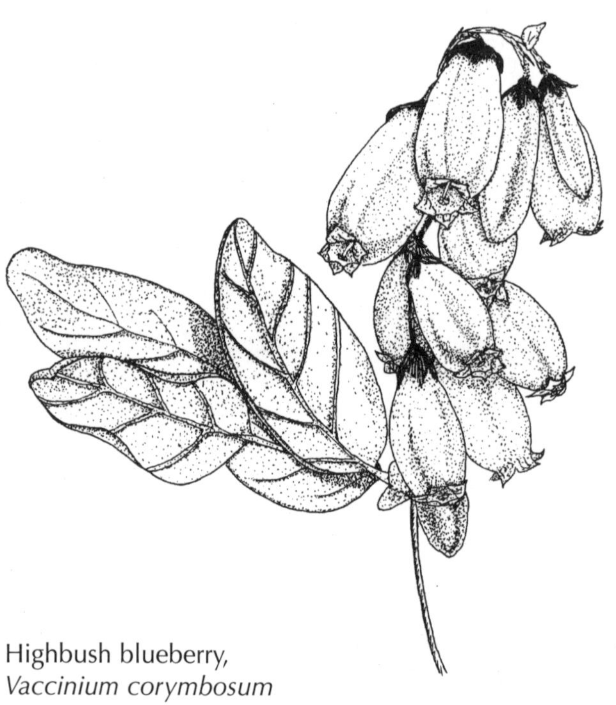

Highbush blueberry,
Vaccinium corymbosum

Application for Certification

School/Learning Center Name ____________________

Contact Name ____________________

Name(s) to Appear on Certificate ____________________

Address ____________________

City ____________ State/Province ______ County ______

Zip/Postal Code ____________ Country ____________

Telephone ____________ E-mail Address ____________

Property Size (Sq. Ft. or Acres) ____________________

Have you ever been certified as a National Wildlife Federation Habitat? ☐ *Yes* ☐ *No*

If yes, list Habitat # ____________________

Office Use:

Habitat # ________

Date Received ________

Certified ________

C.S. ________

Key Words ________

A schoolyard habitat project can be an incredible addition to your school campus and local community for many years to come. After reading the Schoolyard Habitats Planning Guide, please complete all five sections of this application, showing the steps you have taken to create the habitat area and how it is used as a learning site.

You may include photographs of the habitat during development and/or the completed project. We are unable to return materials sent to us, so please be sure to keep duplicates for your records.

I) Project Description/Goals

Please briefly describe current projects and future goals for the schoolyard habitat. Include the process that occurred during the development of your habitat and the educational objectives of this project.

II) Habitat Team Structure

List the key participants involved with the habitat project and the segment of the school community they represent.

Team Member Name: ____________________ Representing: ____________________

III) Habitat Components

A. FOOD/*Plantings and Feeders*

Food can be best provided for wildlife naturally — nuts, nectar, berries, pollen, and buds — or can be made available by bird feeders that may provide food on a year-round basis. Do your best to briefly list the food available for wildlife and if it is provided naturally or offered as a supplement.

Food Type:	Provided By:	Visited By:

B. WATER/*Drinking, Bathing*

We provide water: ☐ Throughout the Year ☐ Seasonally

We provide water in:
- ☐ Bird Bath
- ☐ Pond
- ☐ Spring
- ☐ Water Garden
- ☐ Water Dripping into a Bird Bath
- ☐ Stream
- ☐ Other (describe): ____________________

C. COVER/*Places to Hide*

We provide wind-breaks and places for wildlife to hide from predators in the following manner:

Deciduous Shrubs/Trees:	Number:	Evergreen Shrubs/Trees:	Number:

☐ Brush Piles ☐ Log Piles ☐ Rock Piles/Walls ☐ Ground Covers

☐ Meadow, Scrub, or Prairie Patch ☐ Other (describe): ____________________

D. PLACES TO RAISE YOUNG

We provide the following for nesting birds, denning mammals, egg-laying reptiles and amphibians, fish, butterflies, and other insects and invertebrates.

Large Trees:	Small Trees:	Shrub Masses:

☐ Trees with Nest/Den Cavities ☐ Dens in Ground/Rock

☐ Water Garden/Pond ☐ Nesting Boxes

☐ Nesting Shelves ☐ Meadow, Prairie, or Scrub Patch

☐ Other (describe): ____________________

IV) Site Diagram

Please enclose a sketch or landscape design to visually explain the components of your schoolyard habitat.

V) Curriculum Use

A. Describe how you are using this area for cross-curriculum learning:

B. What resources have been helpful in the planning and instructional use of your Schoolyard Habitats project (e.g., Project WILD, Project LEARNING TREE, NAAEE's VINE program, and Project HOME)?

Thank you **for your interest and participation in the Schoolyard Habitats program.**

- Allow 6-8 weeks for processing of the application.
- Remember to recertify annually.
- Be sure to save copies of all materials you submit as they will not be returned.
- **Please submit the *original copy* of this pre-paid application.**
- Enclose additional pages when necessary.
- Kindly limit your submission to a size no larger than what could fit into the supplied 9" x 12" envelope.

Please send your submission to:

Schoolyard Habitats
National Wildlife Federation
8925 Leesburg Pike
Vienna, VA 22184-0001

Schoolyard Habitats *• National Wildlife Federation • 8925 Leesburg Pike • Vienna, VA 22184-0001*
(703)790-4434 • FAX (703)790-4468 • World Wide Web http://www.nwf.org/nwf/prog/habitats

printed with soy inks on unrebleached recycled paper with 100% post-consumer waste

Glossary

adventitious roots Secondary roots that originate above ground, often in response to flooding.

algal blooms An explosion in the population of algae caused by nutrient-rich water.

alkaline Basic on the pH scale, a pH greater than 7.4.

alluvial Having to do with river processes that deposit sediment on floodplains, deltas, or beds.

anaerobic Living, active, or occurring in the absence of free oxygen. Wetland soils develop under anaerobic conditions due to prolonged flooding or saturation.

anoxic Without free oxygen.

aquifer A layer of permeable rock, sand, or gravel containing groundwater.

animal eatout Consumption of a large area of herbaceous wetland plants by animals.

artesian aquifer An aquifer confined between two layers of less permeable material, resulting in pressure on the groundwater in the aquifer.

basal area The cross sectional area of a tree trunk measured at 4.5 feet (1.4 meters) above ground level; used as a measure of plant species predominance.

base flow The low water level in a perennial stream; that portion of stream flow derived from groundwater.

baseline A known position used to locate something.

benthic organisms Creatures living on or near the bottom of a body of water.

berm A mound or wall of earth.

biological benchmark A plant community (composed of one or more plant species) that has been experiencing the same hydrology over a period of several years, and that will be used as a reference for a planned wetland site.

biomass The amount of organic matter in a given area.

biota All living things in a specific area.

bog A wetland characterized by an accumulation of peat, stagnant water, and high acidity in a nutrient-poor environment. Located in cool, moist, glaciated regions.

bottomland Low-lying relatively flat land of river valleys; land that is subject to periodic flooding.

brackish Containing a mixture of water from rivers and oceans.

buffer zone An area designed to separate wetland inhabitants from human activities.

clay Soil composed of extremely fine, flat particles less than 0.002 mm in size.

community Interacting populations of plants and animals in the same location.

competition An interaction between two organisms or species vying for the same limited resource, whether it be food, shelter, water, or mates.

conifer A cone-bearing tree or shrub, usually evergreen.

contour line A line on a topographic map connecting points of equal elevation.

critical area Areas of special concern such as wetlands, floodplains, steep slopes, and any habitat for rare, threatened, or endangered species.

cyclic dynamics Predictable, often seasonal, ecological interactions within a wetland.

deciduous plant Any plant that sheds its leaves when dormant, as in the cold or dry season.

decomposition The process by which a substance is broken down by bacteria and fungi.

discharge point The location where excess water exits a wetland through a pipe, channel, or culvert.

dormant propagule Part of a plant involved in dispersal and reproduction, during its resting phase.

ecosystem A community of living organisms and its environment, functioning as a unit.

emergent plant Erect, rooted, herbaceous plant with leaves and portion of stem above the water surface; a plant that is tolerant of being completely submerged for short periods only.

erosion The process by which land is worn away by external forces, such as wind, water, or human activity. The loss of soil when carried away by flowing water or wind.

estuarine Having to do with a semi-enclosed bay where river and ocean meet.

evaporation The conversion of liquid water to vapor, the speed of which is increased by higher temperature and wind action.

evapotranspiration The loss of water through both evaporation and transpiration.

evergreen Any plant that keeps its leaves (needles) throughout the year.

facultative plants Plant species in which individuals have a similar likelihood of being found in a wetland or in an upland area; plant species of which 33 to 67% of the population grow in a wetland, such as red maple, bayberry, and big bluestem.

facultative upland plants Plant species in which individuals are more likely to be found in upland areas than wetland areas; plant species of which 1 to 33% of the population grow in a wetland, such as Kentucky bluegrass, eastern hemlock, white oak.

facultative wetland plants Plant species in which individuals are more likely to be found in wetland areas than upland areas; plant species in which 67 to 99% of the population grow in a wetland, such as river birch, jack-in-the-pulpit, and highbush blueberry.

fen An alkaline wetland with an accumulation of peat and abundant nutrient and groundwater inflow.

fetch The distance that waves travel before encountering an obstruction.

floodplain Level land bordering a river or stream, and subject to periodic flooding that causes a buildup of sediment.

gleyed soil Soil in which chemicals have been reduced (oxygen is removed) in an anoxic environment, causing soil colors to appear gray, greenish gray, or bluish gray. It is an indication that soil is, or has been, continuously wet for long periods of time.

grading The process of changing the slope and elevation of land; leveling to a desired horizontal gradient.

grading plan A graphic representation of the prospective topography of an area.

groundwater Water present below the surface in completely saturated soils, sediments, and rocks; water which may flow slowly through the ground.

groundwater discharge Groundwater released via a spring, a seep, bank seepage, or upwelling in a larger body of water; when the water table is above the point of issue (or discharge).

groundwater recharge The downward movement of water through soils (especially in wetlands) to the water table.

growing season The frost-free period of the year.

herbaceous plants Plants with soft non-woody stems that are easily crushed; plants whose above-ground parts die back each year.

herpetology The study of reptiles and amphibians.

hydric soil Soil that is wet long enough during the growing season to produce anaerobic conditions, which are conducive to the growth of hydrophytes.

hydrology The study of water properties, distribution, and circulation.

hydroperiod The length, depth, and frequency of flooding or of near-surface saturation of soils over time. The seasonal or cyclical pattern of flooding or soil saturation.

hydrophyte A plant that grows in water or on a substrate that is periodically deficient in oxygen as a result of excessive water content. A plant usually found in a wetland.

impervious layer A layer that water cannot penetrate.

indicator status A designation that indicates the frequency at which a plant species is found in wetlands (OBL, FACW, FAC, FACU).

infiltration The process by which surface water soaks into the ground; percolation.

intermittent stream A stream channel containing nontidal flowing water part of the year.

intertidal zone An area that is alternately flooded and exposed by tides.

jurisdictional wetlands Wetlands identified by the U. S. Army Corps of engineers as subject to Federal regulation under the Clean Water Act.

lacustrine Pertaining to lakes, reservoirs, and large ponds.

limnetic zone A lake's deep region which lacks rooted vegetation.

littoral zone A lake's shallow region which has nonpersistent emergent vegetation.

loam Soil composed of a mixture of clay, sand, and silt.

lower perennial river River with permanent flowing water and a well-developed floodplain.

macroinvertebrates Animals that do not have backbones and that are large enough to see without magnification.

mangrove forests Tropical wetlands occurring along tidal coastal areas and dominated by mangrove trees, which are characterized by spreading branches that send down roots and form multiple trunks.

marsh A wetland which is characterized by herbaceous emergent vegetation, and flooded permanently or for a major portion of the year.

mean high water (MHW) The average elevation of high water.

mean low water (MLW) The average elevation of low water.

monotypic Pertaining to a plant community consisting of a single species, such as a stand of cattails.

morphology The form and structure of an organism or its parts.

mottling Bright red, orange, or black specks in soil, resulting from oxidation-reduction reactions; iron particles in the soil turn rust red, and manganese particles turn black. Mottling is indicative of periodically wet, but not continuously wet soil conditions.

muck Organic matter so decomposed that very little plant material is identifiable. Muck feels greasy when rubbed between fingers.

nonpersistent emergent plants Plants whose above-ground parts decompose rapidly during dormancy, leaving no evidence of the plant's presence during the dormant season, other than the living roots that remain in the soil.

nontidal regime Hydrology that is not influenced by tides, but instead by surface water runoff and/ or groundwater discharge.

obligate plants Plants adapted to living only in a specific environment, such as a wetland or an upland area.

obligate upland plants Plants especially adapted to living in upland areas. Plant species of which less than 1% of the individuals in the population grow within a wetland, such as cacti.

obligate wetland plants Plants especially adapted to living in wetlands. Plant species of which 99% of the of the individuals in the population grow within a wetland, such as bald cypress and cattail.

palustrine Having to do with small wetlands not associated with rivers or lakes; most inland wetlands, including small ponds.

peat An accumulation of partially decomposed vegetation; organic matter in which the leaves, stems, and roots of plants are still identifiable when gently rubbed between fingers.

persistent emergent plants Plants that remain erect in a marsh even after going dormant for the winter, such as cattails.

pH A measure of the relative acidity of a solution based on a scale of 1 to 14, with 1 being most acidic, 14 being most basic, and 7 being neutral.

photosynthesis The chemical process by which green plants harness energy from the sun to combine carbon dioxide and water, forming a simple sugar (glucose) and oxygen.

piezometer An instrument used to measure hydrostatic pressure. Can be used in bore holes to determine the surface depth (upper boundary) of groundwater.

productivity The rate at which organic matter is produced.

pulsed dynamics Ecological changes resulting from uncommon but recurring events, such as a 100-year storm or animal eatout.

respiration A chemical process that breaks down glucose in the presence of oxygen to release carbon dioxide, water, and energy. The means by which organisms obtain energy from food.

rhizomes A persistent horizontal stem that produces roots and upright stems. If part of the rhizome is broken off, it functions as an independent plant that is genetically identical to the parent plant.

riparian Pertaining to the bank of a river or stream.

riparian forest Bottomland hardwood and floodplain forests along rivers and streams.

riverine Having to do with rivers and streams.

runoff Excess precipitation that cannot be absorbed by soil or vegetation, and that flows into nearby bodies of water, taking sediment with it.

saddle point The land connecting two separate topographic high points.

salinity The salt content of water. The number of grams of dissolved salts in 1,000 grams of water; usually expressed as parts per thousand (ppt).

salt marsh A community of vegetation that is alternately flooded and exposed by tides; found along many coasts. *Spartina* grasses are often the dominant vegetation.

sand Coarse-grained mineral particles 0.05 to 2.0 mm in size.

sediment Fine material that is deposited by water, wind, or glaciers.

seine A large net with weights on the bottom edge and floats on the top edge. Used to enclose fish and other creatures when its ends are drawn together or when the net is pulled ashore.

shrub A woody plant with multiple stems arising near the base; less than 20 feet (6 m) tall at maturity.

silt Fine soil particles 0.002 to 0.05 mm in size.

soil horizon A distinct layer of soil, often parallel to the soil surface and having relatively uniform properties of color, texture, and permeability.

species A category of biological classification, consisting of organisms that have common characteristics and that are potentially capable of interbreeding.

submerged aquatic vegetation (SAV) Plants rooted on the bottom, having leaves, and living entirely underwater.

substrate The base or substance on which species can grow.

subtidal zone Areas continuously submerged under salty or brackish water.

succession The natural change of inhabitants within a community over time.

supratidal zone The area of shoreline above high tide.

swale An open low-lying depression; usually seasonally wet.

swamp A type of wetland characterized by trees and shrubs, and having surface water all or part of the year.

tidal zone An area subject to tidal fluctuations during at least part of the growing season.

topography The shape of the ground surface, including elevation features, and natural and man-made features.

topsoil Surface soil, including the organic layer in which plants have the majority of their roots.

transect A marked line along which measurements are made or vegetation is sampled; usually perpendicular to a baseline.

transit A tool used for surveying; a theodolite with a mounted telescope for siting elevations.

transpiration The process in which water vapor is given off by plants to the atmosphere through the stomata of leaves and stems.

turbidity A measurement of the lack of water clarity. Cloudiness in water results from the scattering of light by algae or other suspended particles.

upper perennial river River with permanent flowing water and very little or no floodplain development.

watershed A region that drains into a specific body of water; a drainage basin.

water table The upper limit of soil saturated by groundwater, and being subject to seasonal variation. In general, the water table is at higher elevation beneath hills and at lower elevation beneath valleys.

weir A dam placed in a stream to raise the water level or divert the water flow.

wetlands Those areas that are inundated or saturated by surface or ground water at a frequency and duration sufficient to support, and that under normal circumstances do support, a prevalence of vegetation typically adapted for life in saturated soil conditions. Wetlands generally include swamps, marshes, bogs, and similar areas.

woody plants Plants with stems protected by bark; trees, shrubs, and some vines.

wrack The accumulation of dead plant debris along a shoreline.

Umbrella sedge,
Cypernus sp.

Appendix A: Activity Selection

The activities in Chapters 8 and 9 are directed towards planning specific types of wetlands. Some activities are appropriate if the goal is to create or restore a wetland. Other activities are directed towards the enhancement of an existing wetland, whereas one activity (*Where Are We?*) is especially useful in locating a wetland to monitor.

Whether creating or restoring a wetland, a preliminary assessment of the available water supply is necessary. Water supplied by oceans, rivers, and large lakes is considered to be unlimited. Water provided by streams, stormwater runoff, or groundwater is limited. Water supplied by a garden hose is artificial.

Select monitoring activities in Chapters 10 and 11, and stewardship activities in Chapter 12 according to the ages of the participants, equipment available, and interests of the group. Activities in these last three chapters are suitable for all types of wetlands.

Marbled salamander, *Ambystoma opacum*

©1997 R.F. Wilson/EC

Chapter	Activity	Page	Creating or Restoring a Wetland			Enhancing a Wetland	Monitoring a Wetland
			Unlimited Water Supply	Limited Water Supply	Artificial Water Supply		
8	Where Are We?	118	X	X	X	X	X
8	What Grows There?	127	X	X	X	X	
8	Making Choices, Setting Goals	137	X	X	X	X	
8	Are We Legal?	144	X	X	X	X	
8	Finding Living Benchmarks	149	X			X	
8	Catch a Flowing Stream	156		X			
8	Will Stormwater Run Off?	163		X			
8	Balancing Water Budgets	169		X	X		
9	Discovering Dips	178	X	X	X		
9	Making the Grade	190	X	X	X		
9	Planting Plans	198	X	X	X	X	
9	Clarifying Specs	208	X	X	X	X	
10	Monitoring a Wetland--all	217					X
11	Watching Wetland Wildlife--all	269					X
12	Wetland Stewardship--all	317	X	X	X	X	X

Appendix B: Correlations with National Standards

Standards-based education began with the publication of the National Science Education Standards developed by the National Research Council. The content standards show which science concepts are essential in order for students to understand our world as we currently know it, allowing educators to focus efforts on the basics before moving on to other topics.

Administrators have seized this idea of prioritizing what is taught, requiring educators to use either the National Science Education Standards, or a state or local version of the standards. This national focus on standards-based education has placed an additional administrative burden on our wetland educators. To ease this situation, we have correlated all *POW!* activities with the national standards.

The correlations are separated into two groups according to grade levels. Correlations for grades 5-8 begin on the next page, and for grades 9-12 on page 336. The correlations read across facing pages with the activities listed in the first column. Filled circles (●) indicate that without using the extensions section of each activity, the activity satisfies a majority of that content standard. Open circles (O) indicate that the activity partially satisfies that content standard.

A more complete description of the National Science Education Standards is available for sale from the National Academy Press, 2101 Constitution Avenue, NW, Box 285, Washington, DC 20055. Call 800-624-6242 or visit www.nas.edu.

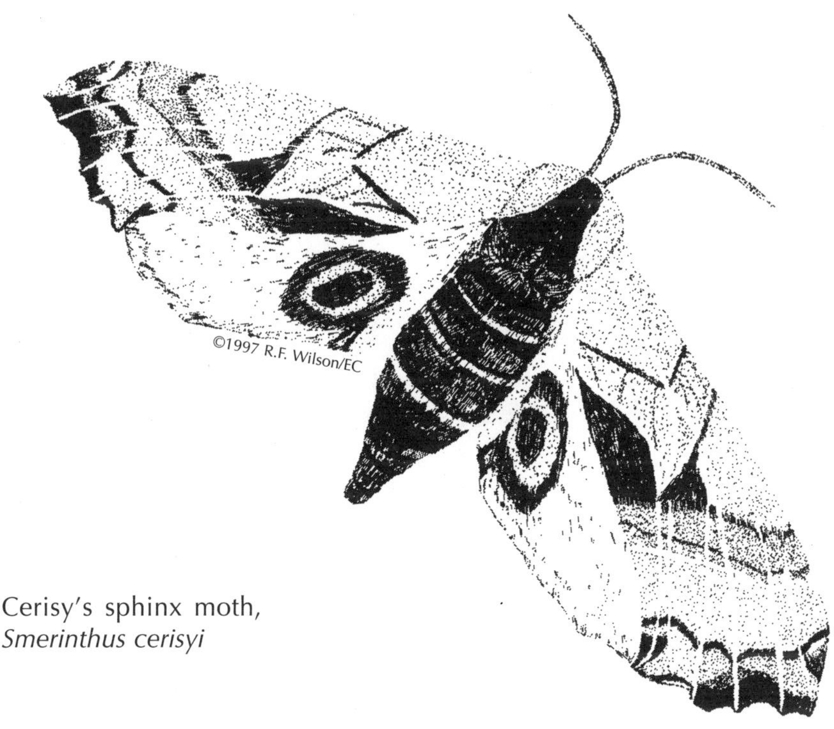

Cerisy's sphinx moth,
Smerinthus cerisyi

CORRELATION OF POW! ACTIVITIES WITH NATIONAL SCIENCE EDUCATION STANDARDS

Grades 5-8

		UNIFYING CONCEPTS & PROCESSES					SCIENCE AS INQUIRY		PHYSICAL SCIENCE			LIFE SCIENCE				
Page	Activity	Systems, order, & organization	Evidence, models, & explanation	Change, constancy, & measurement	Evolution & equilibrium	Form & function	Abilities to do scientific inquiry	Understanding scientific inquiry	Properties & changes of properties	Motions & forces	Transfer of energy	Structure & function in living systems	Reproduction & heredity	Regulation & behavior	Populations & ecosystems	Diversity & adaptations of organisms
118	Where Are We?	●	●	●		●	O	O		O						
127	What Grows There?	●	●	●	●	●	●	O	O			O		●	●	O
137	Making Choices, Setting Goals	O				●									O	
144	Are We Legal?	O														
149	Finding Living Benchmarks	●	O	●	●		O							●		
156	Catch a Flowing Stream	●	O	●	●		O	O		●						
163	Will Stormwater Run Off?	●	●	●			●	O		O						
169	Balancing Water Budgets	●	O				O	O	O							
178	Discovering Dips	O	●	●		●	●	●								
190	Making the Grade	O	●	●		●	●	O								
198	Planting Plans	●				●	●	O						O	●	O
208	Clarifying Specs	O				●										
218	Faithful Flora	●	●	●	●		●	●						O	O	
226	Floral Friends	●	●	●	●		●	●						O	●	
233	Soggy Soils	●	●	●	●	●	●	●	O							
239	Healthy Water	O	●	●	●		●	●			O					
249	Marvelous Hydrology	●	●	●	●	●	●	●								
259	Data Sandwiches	●	●	●	●		●	●	O			O		●	●	
270	Wetland Journeys	O	O			O	O	O						O		
273	Mammal Marks	O	●			●	O	O						O	O	
281	Beautiful Birds	O	●	●		O	O	O						O	O	
288	Herp Search	●	●	●		●	O	O						O	O	
296	Bugs and Butterflies	●	●			●	O	O						O	O	
303	Aquatic Animals	●	●	●		O	O	O						O	O	
310	Wetland Soup	●	●	●		●	O	O						O	O	

CORRELATION OF POW! ACTIVITIES WITH NATIONAL SCIENCE EDUCATION STANDARDS

Grades 5-8

Page	Activity	EARTH & SPACE SCIENCE			SCIENCE & TECHNOLOGY		SCIENCE IN PERSONAL & SOCIAL PERSPECTIVES					HISTORY & NATURE OF SCIENCE		
		Structure of the earth system	Earth's history	Earth in the solar system	Abilities of technological design	Understanding science & technology	Personal health	Populations, resources, environments	Natural hazards	Risks & benefits	Science & technology in society	Science as a human endeavor	Nature of science	History of science
118	Where Are We?					O								
127	What Grows There?	O										O	O	
137	Making Choices, Setting Goals				O		O		O	O	O			
144	Are We Legal?	O								O	O	O		
149	Finding Living Benchmarks											O		
156	Catch a Flowing Stream	O					O					O		
163	Will Stormwater Run Off?	O										O		
169	Balancing Water Budgets	O		O	O	O					O	O	O	
178	Discovering Dips				O	O					O	O	O	
190	Making the Grade				●	●					O	O	O	
198	Planting Plans				●	●					O	O	O	
208	Clarifying Specs					O	O		O	O		O		
218	Faithful Flora							●	O			O	O	
226	Floral Friends							●	O			O	O	
233	Soggy Soils	O						●	O			O	O	
239	Healthy Water	O				O	O	●	O	O		O	O	
249	Marvelous Hydrology	O			O	O		●	O		O	O	O	
259	Data Sandwiches	O			O	O		●	O		O	O	O	
270	Wetland Journeys											O	O	O
273	Mammal Marks											O	O	
281	Beautiful Birds											O	O	
288	Herp Search											O	O	
296	Bugs and Butterflies											O	O	
303	Aquatic Animals											O	O	
310	Wetland Soup						O	●	O	O		O	O	

CORRELATION OF POW! ACTIVITIES WITH NATIONAL SCIENCE EDUCATION STANDARDS

Grades 9-12

Page	Activity	UNIFYING CONCEPTS & PROCESSES					SCIENCE AS INQUIRY		PHYSICAL SCIENCE						LIFE SCIENCE					
		Systems, order, & organization	Evidence, models, & explanation	Change, constancy, & measurement	Evolution & equilibrium	Form & function	Abilities to do scientific inquiry	Understanding scientific inquiry	Structure of atoms	Structure & properties of matter	Chemical reactions	Motions & forces	Conservation of energy	Interactions of energy & matter	The cell	Molecular basis of heredity	Biological evolution	Interdependence of organisms	Matter, energy, & organization	Behavior of organisms
118	Where Are We?	●	●	●		●														
127	What Grows There?	●	●	●	●	●	O	O			O							O	O	O
137	Making Choices, Setting Goals	O				●												O	O	
144	Are We Legal?	O																O		
149	Finding Living Benchmarks	●	O	●	●		O	O									O	O	O	O
156	Catch a Flowing Stream	●	O	●	●		O	O												
163	Will Stormwater Run Off?	●	●	●			●	O												
169	Balancing Water Budgets	●	O				O	O												
178	Discovering Dips	O	●	●		●	●	●												
190	Making the Grade	O	●	●		●	●	●												
198	Planting Plans	●				●	●	●										O	O	
208	Clarifying Specs	O				●														
218	Faithful Flora	●	●	●	●		●	●										O	O	O
226	Floral Friends	●	●	●	●		●	●										O	O	O
233	Soggy Soils	●	●	●	●	●	●	●			O									
239	Healthy Water	O	●	●	●		●	●			O									
249	Marvelous Hydrology	●	●	●	●	●	●	●												
259	Data Sandwiches	●	●	●	●		●	●			O							O	O	
270	Wetland Journeys	O	O			O	O	O												
273	Mammal Marks	O	●			●	O	O										O	O	O
281	Beautiful Birds	O	●	●		O	O	O										O	O	O
288	Herp Search	●	●	●		●	O	O										O	O	O
296	Bugs and Butterflies	●	●			●	O	O										O	O	O
303	Aquatic Animals	●	●	●		O	O	O										O	O	O
310	Wetland Soup	●	●	●		●	O	O										O	O	O

CORRELATION OF POW! ACTIVITIES WITH NATIONAL SCIENCE EDUCATION STANDARDS

Grades 9-12

Page	Activity	EARTH & SPACE SCIENCE				SCIENCE & TECHNOLOGY		SCIENCE IN PERSONAL & SOCIAL PERSPECTIVES						HISTORY & NATURE OF SCIENCE		
		Energy in the earth system	Geochemical cycles	Origin & evolution of earth system	Origin & evolution of universe	Abilities of technological design	Understanding science & technology	Personal & community health	Population growth	Natural resources	Environmental quality	Natural & human-induced hazards	Science & technology challenges	Science as a human endeavor	Nature of scientific knowledge	Historical perspectives
118	Where Are We?					O	O									
127	What Grows There?													O		
137	Making Choices, Setting Goals							O		O	O	O	O			
144	Are We Legal?									O	O	O				
149	Finding Living Benchmarks													O		
156	Catch a Flowing Stream													O		
163	Will Stormwater Run Off?													O		
169	Balancing Water Budgets	O	O			O	O			O				O		
178	Discovering Dips					●	O							O		
190	Making the Grade					●	O							O		
198	Planting Plans					●	O							O		
208	Clarifying Specs					O	O	O				O		O		
218	Faithful Flora									O	O			O	●	O
226	Floral Friends									O	O			O	●	O
233	Soggy Soils									O	O		O	O	●	O
239	Healthy Water							O		O	O	O	O	O	●	O
249	Marvelous Hydrology		O			O	O			O	O		O	O	●	O
259	Data Sandwiches					O	O				O		O	O	●	O
270	Wetland Journeys													O	O	
273	Mammal Marks													O	O	
281	Beautiful Birds													O	O	
288	Herp Search													O	O	
296	Bugs and Butterflies													O	O	
303	Aquatic Animals													O	O	
310	Wetland Soup							O		O	O	O	O	O	O	

Appendix C: Examples of Planned Wetlands

Throughout Chapter 9 data and plans are shown for a planned wetland project at Horsehead Wetlands Center located in Grasonville, Maryland. The data were produced by educators attending a pilot *POW!* workshop at the site. As they worked through the activities from *POW!*, these educators designed the wetland. The constraints on water supply, location, and size of this planned wetland are similar to many schoolyard situations.

The site provided for the project was located adjacent to a pavilion used for educational activities and bounded by a gravel drive and the building. The workshop participants determined that runoff from one side of the pavilion roof would be adequate for a small wetland (approximately 10 feet by 18 feet) with a center depth of one foot. Well water and a hose were available to provide wetland plants with additional moisture during the first year while they were becoming established and as a backup water source should a drought occur (as did happen the first summer).

The photographs that follow show changes that have occurred in the wetland during its first eighteen months. Hopefully others will be encouraged to take on the rewarding task of planning a schoolyard wetland.

Figure A.1 The location for the planned wetland was bounded by a gravel driveway on the south and east sides and an educational pavilion on the west northwest side. (See Figure 9.9 on page 200 for more complete orientation.) Rainwater falling on half of the pavilion roof was directed from two downspouts through four-inch flex pipe into the planned wetland. (A swale was planned originally, but changed to buried flex pipe which was less of a tripping hazard.) The soil was clay loam, therefore neither soil compaction nor a liner was needed.

Figure A.2 Survey flags and spray paint marked the area to be dug. Excavated soils were used to create a berm in a low area between the wetland and the gravel driveway.

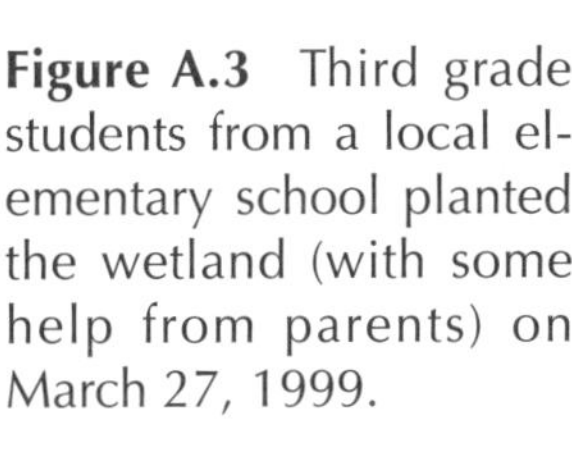

Figure A.3 Third grade students from a local elementary school planted the wetland (with some help from parents) on March 27, 1999.

Figure A.4 Students celebrated a successful planting and raised their feet as the wetland filled for the first time with water supplied by a hose.

Figure A.5 By April 22, 1999, many of the plants had been eaten by waterfowl, turkeys, rabbits, and deer. The wetland plants that had been grown from seeds in a greenhouse during late winter, were viewed by wildlife as a fresh salad compared to surrounding vegetation. Some plant species were ignored, but many were eaten. The viewer is facing west.

Figure A.6 Some of the educators who planned the wetland practice monitoring techniques for water, soils, and plants during a workshop in April 1999.

Figure A. 7 By July 21, 1999, the surviving plants were dealing with a major regional drought. The viewer is facing west.

Figure A.8 The surviving plants received extra water from the pump visible at the edge of the building. The viewer is facing north in this July 21st photograph.

Figure A.9 By August 10, 2000, the wetland was thriving. Some species had been replanted. A split rail fence was installed to deter the deer and large mesh wire was attached to the lower rail to prevent waterfowl from walking into the wetland. The viewer is facing west.

Figure A.10 Most plants were blooming in August 2000, and the wetland was busy with butterflies, frogs, and turtles. The viewer is facing north.

Figure A.11 By September 23, 2000, the wetland vegetation is lush and alive with wildlife. The viewer is facing west in this close-up view of the wetland.

Figure A.12 During the second summer no extra water was added; the wetland survived on roof runoff alone. Herbivory had been limited, but not eliminated. As of September 2000, the wetland appears to be successful. The viewer is facing north.

Index